# Advances in Intelligent and Soft Computing 63

Editor-in-Chief: J. Kacprzyk

# Advances in Intelligent and Soft Computing

**Editor-in-Chief**

Prof. Janusz Kacprzyk
Systems Research Institute
Polish Academy of Sciences
ul. Newelska 6
01-447 Warsaw
Poland
E-mail: kacprzyk@ibspan.waw.pl

Further volumes of this series can be found on our homepage: springer.com

Vol. 47. E. Pietka, J. Kawa (Eds.)
*Information Technologies in Biomedicine*, 2008
ISBN 978-3-540-68167-0

Vol. 48. D. Dubois, M. Asunción Lubiano,
H. Prade, M. Ángeles Gil,
P. Grzegorzewski,
O. Hryniewicz (Eds.)
*Soft Methods for Handling
Variability and Imprecision*, 2008
ISBN 978-3-540-85026-7

Vol. 49. J.M. Corchado, F. de Paz,
M.P. Rocha,
F. Fernández Riverola (Eds.)
*2nd International Workshop on Practical
Applications of Computational Biology
and Bioinformatics (IWPACBB 2008)*, 2009
ISBN 978-3-540-85860-7

Vol. 50. J.M. Corchado, S. Rodriguez,
J. Llinas, J.M. Molina (Eds.)
*International Symposium on Distributed
Computing and Artificial Intelligence 2008
(DCAI 2008)*, 2009
ISBN 978-3-540-85862-1

Vol. 51. J.M. Corchado, D.I. Tapia,
J. Bravo (Eds.)
*3rd Symposium of Ubiquitous
Computing and Ambient
Intelligence 2008*, 2009
ISBN 978-3-540-85866-9

Vol. 52. E. Avineri, M. Köppen,
K. Dahal,
Y. Sunitiyoso, R. Roy (Eds.)
*Applications of Soft Computing*, 2009
ISBN 978-3-540-88078-3

Vol. 53. E. Corchado, R. Zunino,
P. Gastaldo, Á. Herrero (Eds.)
*Proceedings of the International
Workshop on Computational
Intelligence in Security for
Information Systems CISIS 2008*, 2009
ISBN 978-3-540-88180-3

Vol. 54. B.-y. Cao, C.-y. Zhang,
T.-f. Li (Eds.)
*Fuzzy Information and Engineering*, 2009
ISBN 978-3-540-88913-7

Vol. 55. Y. Demazeau, J. Pavón,
J.M. Corchado, J. Bajo (Eds.)
*7th International Conference on Practical
Applications of Agents and Multi-Agent
Systems (PAAMS 2009)*, 2009
ISBN 978-3-642-00486-5

Vol. 56. H. Wang, Y. Shen,
T. Huang, Z. Zeng (Eds.)
*The Sixth International Symposium on Neural
Networks (ISNN 2009)*, 2009
ISBN 978-3-642-01215-0

Vol. 57. M. Kurzynski,
M. Wozniak (Eds.)
*Computer Recognition Systems 3*, 2009
ISBN 978-3-540-93904-7

Vol. 58. J. Mehnen, A. Tiwari,
M. Köppen, A. Saad (Eds.)
*Applications of Soft Computing*, 2009
ISBN 978-3-540-89618-0

Vol. 59. K.A. Cyran,
S. Kozielski, J.F. Peters,
U. Stańczyk, A. Wakulicz-Deja (Eds.)
*Man-Machine Interactions*, 2009
ISBN 978-3-642-00562-6

Vol. 60. Z.S. Hippe,
J.L. Kulikowski (Eds.)
*Human-Computer Systems Interaction*, 2009
ISBN 978-3-642-03201-1

Vol. 61. XXX

Vol. 62. Bingyuan Cao,
Tai-Fu Li, Cheng-Yi Zhang (Eds.)
*Fuzzy Information and Engineering Volume 2*, 2009
ISBN 978-3-642-03663-7

Vol. 63. Álvaro Herrero, Paolo Gastaldo,
Rodolfo Zunino, and Emilio Corchado (Eds.)
*Computational Intelligence in Security for
Information Systems*, 2009
ISBN 978-3-642-04090-0

Álvaro Herrero, Paolo Gastaldo,
Rodolfo Zunino, and Emilio Corchado (Eds.)

# Computational Intelligence in Security for Information Systems

CISIS'09, 2nd International Workshop
Burgos, Spain, September 2009 Proceedings

**Editors**

Álvaro Herrero
Grupo de Inteligencia
Computacional Aplicada
Área de Lenguajes y
Sistemas Informáticos
Escuela Politécnica Superior
Universidad de Burgos
Calle Francisco de Vitoria S/N,
Edifico C
09006, Burgos
Spain
E-mail: ahcosio@ubu.es

Paolo Gastaldo
Dept. of Biophysical and
Electronic Engineering
Genova University
Via Opera Pia 11a
16145 Genova
Italy
E-mail: paolo.gastaldo@unige.it

Rodolfo Zunino
Dept. of Biophysical and
Electronic Engineering
Genova University
Via Opera Pia 11a
16145 Genova
Italy
E-mail: rodolfo.zunino@unige.it

Emilio Corchado
Grupo de Inteligencia
Computacional Aplicada
Área de Lenguajes y Sistemas Informáticos
Escuela Politénica Superior
Universidad de Burgos
Calle Francisco de Vitoria S/N,
Edifico C
09006, Burgos
Spain
E-mail: escorchado@ubu.es

ISBN 978-3-642-04090-0            e-ISBN 978-3-642-04091-7

DOI 10.1007/978-3-642-04091-7

Advances in Intelligent and Soft Computing            ISSN 1867-5662

Library of Congress Control Number: 2009933686

©2009 Springer-Verlag Berlin Heidelberg

This work is subject to copyright. All rights are reserved, whether the whole or part of the material is concerned, specifically the rights of translation, reprinting, reuse of illustrations, recitation, broadcasting, reproduction on microfilm or in any other way, and storage in data banks. Duplication of this publication or parts thereof is permitted only under the provisions of the German Copyright Law of September 9, 1965, in its current version, and permission for use must always be obtained from Springer. Violations are liable for prosecution under the German Copyright Law.

The use of general descriptive names, registered names, trademarks, etc. in this publication does not imply, even in the absence of a specific statement, that such names are exempt from the relevant protective laws and regulations and therefore free for general use.

*Typeset & Cover Design:* Scientific Publishing Services Pvt. Ltd., Chennai, India.

Printed in acid-free paper

5 4 3 2 1 0

springer.com

# Preface

The Second International Workshop on Computational Intelligence for Security in Information Systems (CISIS'09) presented the most recent developments in the dynamically expanding realm of several fields such as Data Mining and Intelligence, Infrastructure Protection, Network Security, Biometry and Industrial Perspectives.

The International Workshop on Computational Intelligence for Security in Information Systems (CISIS) proposes a forum to the different communities related to the field of intelligent systems for security. The global purpose of CISIS conferences has been to form a broad and interdisciplinary meeting ground offering the opportunity to interact with the leading industries actively involved in the critical area of security, and have a picture of the current solutions adopted in practical domains.

This volume of Advances in Intelligent and Soft Computing contains accepted papers presented at CISIS'09, which was held in Burgos, Spain, on September $23^{rd}$-$26^{th}$, 2009. After a through peer-review process, the International Program Committee selected 25 papers which are published in this workshop proceedings. This allowed the Scientific Committee to verify the vital and crucial nature of the topics involved in the event, and resulted in an acceptance rate close to 50% of the originally submitted manuscripts.

As a follow-up of the conference, we anticipate further publication of selected papers in a special issue of the Journal of Information Assurance and Security (JIAS). The extended papers, together with contributed articles received in response to subsequent open calls, will go through further rounds of peer refereeing in the remits of this journal.

We would like to thank the work of the Programme Committee Members who performed admirably under tight deadline pressures. Our warmest and special thanks go to the Keynote Speaker: Prof. Bogdan Gabrys from Bournemouth University, UK.

Particular thanks go as well to the Workshop main Sponsors, Junta de Castilla y León, University of Burgos, Diputación de Burgos, Ayuntamiento de Burgos, GCI, CSA, FAE and FEC, who jointly contributed in an active and constructive manner to the success of this initiative.

We wish to thank Prof. Dr. Janusz Kacprzyk (Editor-in-chief), Dr. Thomas Ditzinger (Senior Editor, Engineering/Applied Sciences) and Mrs. Heather King at Springer-Verlag for their help and collaboration in this demanding scientific publication project.

We thank as well all the authors and participants for their great contributions that made this conference possible and all the hard work worthwhile.

July 2009

Álvaro Herrero
Paolo Gastaldo
Emilio Corchado
Rodolfo Zunino

# Organization

## Honorary Chairs

Vicente Orden — Presidente de la Diputación provincial de Burgos (Spain)
Juan Carlos Aparicio — Alcalde de la ciudad de Burgos (Spain)

## General Chairs

Emilio Corchado — University of Burgos (Spain)
Rodolfo Zunino — University of Genoa (Italy)

## Program Committee

### Chairs

Bruno Baruque — University of Burgos (Spain)
Emilio Corchado — University of Burgos (Spain)
Paolo Gastaldo — University of Genoa (Spain)
Álvaro Herrero — University of Burgos (Spain)
Rodolfo Zunino — University of Genoa (Italy)

### Members

Policarpo Abascal Fuentes — University of Oviedo (Spain)
Emilio Aced Félez — Spanish Agency for Data Protection (Spain)
Santiago Martín — Católica del Ecuador Pontificial University (Ecuador)
Isaac Agudo Ruiz — University of Malaga (Spain)
Salvador Alcaraz Carrasco — Miguel Hernández University (Spain)
Cristina Alcaraz Tello — University of Malaga (Spain)
Cesare Alippi — Politecnico di Milano (Italy)
Gonzalo Alvarez Marañón — CSIC (Spain)
Wilson Andia Cuiza — Franz Tamayo University (Bolivia)
Davide Anguita — University of Genoa (Italy)
Enrico Appiani — Elsag Datamat (Italy)
Jhon César Arango Serna — University of Manizales (Colombia)
Javier Areitio Bertolín — University of Deusto (Spain)
Alessandro Armando — University of Genoa (Italy)
Ángel Arroyo — University of Burgos (Spain)
Juan Jesús Barbarán Sánchez — University of Granada (Spain)
Carlos Blanco Bueno — University of Castilla la Mancha (Spain)

Piero Bonissone — GE Global Research (USA)
Miguel Angel Borges Trenard — Oriente University (Cuba)
Antoni Bosch Pujol — Autonomous University of Barcelona (Spain)
Andrés Bustillo — University of Burgos (Spain)
Pino Caballero Gil — University of La Laguna (Spain)
Sandra Patricia Camacho — Javeriana Pontifical University (Colombia)
Juan Carlos Canavelli — Autonomous University of Entre Ríos (Argentina)
Javier Carbó Rubiera — University Carlos III of Madrid (Spain)
Eduardo Carozo Blumsztein — University of Montevideo (Uruguay)
Javier Fernando Castaño — University of los Llanos (Colombia)
Joan-Josep Climent — University of Alicante (Spain)
Rodrigo Adolfo Cofré Loyola — Católica del Maule University (Chile)
Ricardo Contreras Arriagada — University of Concepción (Chile)
Juan Manuel Corchado — University of Salamanca (Spain)
Rafael Corchuelo — University of Seville (Spain)
Andre CPLF de Carvalho — University of São Paulo (Brazil)
Leticia Curiel — University of Burgos (Spain)
Keshav Dahal — University of Bradford (UK)
Enrique Daltabuit — National Autonomous of México University (México)
Enrique De la Hoz de la Hoz — University of Alcalá (Spain)
Sergio Decherchi — University of Genoa (Italy)
Gabriel Díaz Orueta — UNED (Spain)
José A. Domínguez Pérez — University of Salamanca (Spain)
José Dorronsoro — Autonomous University of Madrid (Spain)
María Isabel Dorta González — University of La Laguna (Spain)
Mario Farias-Elinos — La Salle University (Mexico)
Bianca Falcidieno — CNR (Italy)
Dario Forte — University of Milano Crema (Italy)
Amparo Fúster Sabater — CSIC (Spain)
Bogdan Gabrys — Bournemouth University (UK)
Luis Javier García Villalba — Complutense de Madrid University (Spain)
Joaquín García-Alfaro — Carleton University (Canada)
Marcos Gestal Pose — University of La Coruña (Spain)
Juan Manuel González Nieto — Queensland University of Technology (Australia)
Petro Gopych — V.N. Karazin Kharkiv National University (Ukraine)
Manuel Graña — University of Pais Vasco (Spain)
Angel Grediaga Olivo — University of Alicante (Spain)
Juan Pedro Hecht — University of Buenos Aires (Argentina)
Julio Cesar Hernandez Castro — University of Portsmouth (UK)

Luis Hernández Encinas — CSIC (Spain)
Candelaria Hernández Goya — University of La Laguna (Spain)
Francisco Herrera — University of Granada (Spain)
R. J. Howlett — University of Brighton (UK)
Llorenç Huguet Rotger — Illes Ballears University (Spain)
José Luis Imaña — Complutense de Madrid University (Spain)
Giacomo Indiveri — ETH Zurich (Switzerland)
Gustavo Adolfo Isaza — University of Caldas (Colombia)
Lakhmi Jain — University of South Australia (Australia)
Janusz Kacprzyk — Polish Academy of Sciences (Poland)
Juha Karhunen — Helsinki University of Technology (Finland)
Juan Guillermo Lalinde-Pulido — EAFIT University (Colombia)
Alain Lamadrid Vallina — Department of High Education (Cuba)
Davide Leoncini — University of Genoa (Italy)
Ricardo Llamosa-Villalba — Industrial University of Santander (Colombia)
Jorge López — University Carlos III of Madrid (Spain)
María Victoria López — Complutense de Madrid University (Spain)
Gabriel López — University of Murcia (Spain)
Antonio Lioy — Politecnico di Torino (Italy)
Wenjian Luo — University of Science and Technology of China (China)
Nadia Mazzino — Ansaldo STS (Italy)
José Francisco Martínez — INAOE (Mexico)
Ermete Meda — Ansaldo STS (Italy)
Evangelia Micheli-Tzanakou — Rutgers University (USA)
Josep M. Miret Biosca — University of Lleida (Spain)
José Manuel Molina — University Carlos III of Madrid (Spain)
Francisco José Navarro Ríos — University of Granada (Spain)
Gianluca Maiolini — University of Rome La Sapienza (Italy)
Constantino Malagón Luque — University Antonio de Nebrija (Spain)
Chelo Malagón Poyato — CSIC (Spain)
Edwin Mamani Zeballos — Military School of Engineering (Bolivia)
Paul Mantilla — Católica del Ecuador Pontifical University (Ecuador)
Antonio Maña Gómez — University of Malaga (Spain)
Angel Martín del Rey — University of Salamanca (Spain)
Rafael Martínez Gasca — University of Seville (Spain)
Gonzalo Martínez Ginesta — Alfonso X El Sabio University (Spain)
Edgar Martínez Moro — University of Valladolid (Spain)
Leandro Mascarello — National University of la Matanza (Argentina)
Mario Mastriani — LIDeNTec (Argentina)
Reinaldo Nicolás Mayol — University of Los Andes (Venezuela)
F. Rolando Menchaca — National Polytechnic Institute (Mexico)
Carlos Mex Perera — University of Monterrey (Mexico)

| | |
|---|---|
| José Antonio Montenegro | University of Malaga (Spain) |
| Guillermo Morales-Luna | CINVESTAV (Mexico) |
| Ivonne Valeria Muñoz | ITESM (Mexico) |
| Macià Mut Puigserver | Illes Ballears University (Spain) |
| Dennis K Nilsson | Chalmers University of Technology (Sweden) |
| Yulier Nuñez Musa | José Antonio Echeverría Polytechnic High Institute (Cuba) |
| Tomas Olovsson | Chalmers University of Technology (Sweden) |
| José Antonio Onieva | University of Malaga (Spain) |
| Juan José Ortega Daza | University of Malaga (Spain) |
| Amalia Beatriz Orúe | CSIC (Spain) |
| Hugo Pagola | University of Buenos Aires (Argentina) |
| Rosaura Palma Orozco | CINVESTAV IPN (México) |
| Danilo Pástor Ramírez | Polytechnic High School of Chimborazo (Ecuador) |
| Mª Magdalena Payeras | Illes Ballears University (Spain) |
| Alberto Peinado | University of Malaga (Spain) |
| Witold Pedrycz | University of Alberta (Canada) |
| Carlos Pereira | University of Coimbra (Portugal) |
| Kostas Plataniotis | University of Toronto (Canada) |
| Francisco Plaza | University of Salamanca (Spain) |
| Pedro Pablo Pinacho | University of Santiago de Chile (Chile) |
| Fernando Podio | NIST (USA) |
| Marios Polycarpou | University of Cyprus (Cyprus) |
| Jorge Posada | VICOMTech (Spain) |
| Sergio Pozo | University of Seville (Spain) |
| Araceli Queiruga | University of Salamanca (Spain) |
| Benjamín Ramos | University Carlos III of Madrid (Spain) |
| Judith Redi | University of Genoa (Italy) |
| Raquel Redondo | University of Burgos (Spain) |
| Perfecto Reguera | University of Leon (Spain) |
| Verónica Requena | University of Alicante (Spain) |
| Arturo Ribagorda | University Carlos III of Madrid (Spain) |
| Bernardete Ribeiro | University of Coimbra (Portugal) |
| Sandro Ridella | University of Genoa (Italy) |
| José Luis Rivas López | University of Vigo (Spain) |
| Ramón Rizo | University of Alicante (Spain) |
| Sergi Robles Martínez | Autonomous University of Barcelona (Spain) |
| Lídice Romero Amondaray | Oriente University (Cuba) |
| Paulo Andrés Ruiz-Tagle | University of Talca (Chile) |
| Dymirt Ruta | British Telecom (UK) |
| José Esteban Saavedra | University of Oruro (Bolivia) |
| José Luis Salazar | University of Zaragoza (Spain) |
| Jose L. Salmeron | Pablo Olavide University (España) |
| Pedro Santos | University of Burgos (Spain) |
| Javier Sedano | University of Burgos (Spain) |
| Roberto Sepúlveda Lima | José Antonio Echeverría Polytechnic High Institute (Cuba) |

| | |
|---|---|
| Manuel Angel Serrano | University of Castilla la Mancha (Spain) |
| Fabio Scotti | University of Milan (Italy) |
| Kate Smith-Miles | Deakin University (Australia) |
| Sorin Stratulat | University Paul Verlaine, Metz (France) |
| Jorge Eduardo Sznek | Nacional del Comahue University (Argentina) |
| Juan Tapiador | University of York (UK) |
| Juan Tena Ayuso | University of Valladolid (Spain) |
| Antonio J. Tomeu | University of Cadiz (Spain) |
| Ramón Torres Rojas | Marta Abreu de Las Villas Central University (Cuba) |
| Leandro Tortosa Grau | University of Alicante (Spain) |
| Fernando Tricas García | University of Zaragoza (Spain) |
| Carmela Troncoso | Katholieke University of Leuven (Belgium) |
| Roberto Uribeetxeberria | University of Mondragon (Spain) |
| Belén Vaquerizo | University of Burgos (Spain) |
| Fabián Velásquez Clavijo | University of los Llanos (Colombia) |
| José Francisco Vicent Francés | University of Alicante (Spain) |
| Tzai-Der Wang | Cheng Shiu University (Taiwan) |
| Lei Xu | Chinese University of Hong Kong (Hong Kong) |
| Xin Yao | University of Birmingham (UK) |
| Hujun Yin | University of Manchester (UK) |
| Alessandro Zanasi | TEMIS (France) |
| David Zhang | Hong Kong Polytechnic University (Hong Kong) |
| Urko Zurutuza | University of Mondragon (Spain) |

## Organizing Committee

**Chairs**

| | |
|---|---|
| Bruno Baruque | University of Burgos (Spain) |
| Emilio Corchado | University of Burgos (Spain) |
| Álvaro Herrero | University of Burgos (Spain) |

**Members**

| | |
|---|---|
| Ángel Arroyo | University of Burgos (Spain) |
| Andrés Bustillo | University of Burgos (Spain) |
| Leticia Curiel | University of Burgos (Spain) |
| Sergio Decherchi | University of Genoa (Italy) |
| Paolo Gastaldo | University of Genoa (Italy) |
| Francesco Picasso | University of Genoa (Italy) |
| Judith Redi | University of Genoa (Italy) |
| Raquel Redondo | University of Burgos (Spain) |
| Pedro Santos | University of Burgos (Spain) |
| Javier Sedano | University of Burgos (Spain) |
| Belén Vaquerizo | University of Burgos (Spain) |

# Table of Contents

## Data Mining and Intelligence

A Data Mining Based Analysis of Nmap Operating System Fingerprint Database .................................................................. 1
    *João Paulo S. Medeiros, Agostinho M. Brito Jr., and Paulo S. Motta Pires*

Knowledge System for Application of Computer Security Rules ........ 9
    *Menchaca García Felipe Rolando and Contreras Hernández Salvador*

Clustering of Windows Security Events by Means of Frequent Pattern Mining .......................................................................... 19
    *Rosa Basagoiti, Urko Zurutuza, Asier Aztiria, Guzmán Santafé, and Mario Reyes*

Text Clustering for Digital Forensics Analysis ....................... 29
    *Sergio Decherchi, Simone Tacconi, Judith Redi, Alessio Leoncini, Fabio Sangiacomo, and Rodolfo Zunino*

## Infrastructure Protection

A Preliminary Study on SVM Based Analysis of Underwater Magnetic Signals for Port Protection ........................................... 37
    *Davide Leoncini, Sergio Decherchi, Osvaldo Faggioni, Paolo Gastaldo, Maurizio Soldani, and Rodolfo Zunino*

Fuzzy Rule Based Intelligent Security and Fire Detector System ....... 45
    *Joydeb Roy Choudhury, Tribeni Prasad Banerjee, Swagatam Das, Ajith Abraham, and Václav Snášel*

A Scaled Test Bench for Vanets with RFID Signalling ................ 53
    *Andrés Ortiz, Alberto Peinado, and Jorge Munilla*

A SVM-Based Behavior Monitoring Algorithm towards Detection of Un-desired Events in Critical Infrastructures ....................... 61
    *Y. Jiang, J. Jiang, and P. Capodieci*

## Network Security

Design and Implementation of High Performance Viterbi Decoder for Mobile Communication Data Security .............................. 69
    *T. Menakadevi and M. Madheswaran*

An Adaptive Multi-agent Solution to Detect DoS Attack in SOAP
Messages.................................................... 77
  Cristian I. Pinzón, Juan F. De Paz, Javier Bajo, and
  Juan M. Corchado

A Self-learning Anomaly-Based Web Application Firewall............. 85
  Carmen Torrano-Gimenez, Alejandro Perez-Villegas, and
  Gonzalo Alvarez

An Investigation of Multi-objective Genetic Algorithms for Encrypted
Traffic Identification............................................ 93
  Carlos Bacquet, A. Nur Zincir-Heywood, and Malcolm I. Heywood

A Multi-objective Optimisation Approach to IDS Sensor Placement.... 101
  Hao Chen, John A. Clark, Juan E. Tapiador, Siraj A. Shaikh,
  Howard Chivers, and Philip Nobles

Towards Ontology-Based Intelligent Model for Intrusion Detection and
Prevention....................................................... 109
  Gustavo Isaza, Andrés Castillo, Manuel López, and Luis Castillo

Ontology-Based Policy Translation................................ 117
  Cataldo Basile, Antonio Lioy, Salvatore Scozzi, and Marco Vallini

Automatic Rule Generation Based on Genetic Programming for Event
Correlation...................................................... 127
  G. Suarez-Tangil, E. Palomar, J.M. de Fuentes, J. Blasco, and
  A. Ribagorda

Learning Program Behavior for Run-Time Software Assurance......... 135
  Hira Agrawal, Clifford Behrens, Balakrishnan Dasarathy, and
  Leslie Lee Fook

Multiagent Systems for Network Intrusion Detection: A Review...... 143
  Álvaro Herrero and Emilio Corchado

# Biometry

Multimodal Biometrics: Topics in Score Fusion ..................... 155
  Luis Puente, M. Jesús Poza, Juan Miguel Gómez, and Diego Carrero

Security Efficiency Analysis of a Biometric Fuzzy Extractor for Iris
Templates........................................................ 163
  F. Hernández Álvarez and L. Hernández Encinas

Behavioural Biometrics Hardware Based on Bioinformatics Matching... 171
  Slobodan Bojanić, Vukašin Pejović, Gabriel Caffarena,
  Vladimir Milovanović, Carlos Carreras, and Jelena Popović

## Industrial Perspectives

Robust Real-Time Face Tracking Using an Active Camera ............ 179
    *Paramveer S. Dhillon*

An Approach to Centralized Control Systems Based on Cellular
Automata....................................................... 187
    *Rosaura Palma-Orozco, Gisela Palma-Orozco,*
    *José de Jesús Medel-Juárez, and José Alfredo Jiménez-Benítez*

Intelligent Methods and Models in Transportation.................... 193
    *M$^a$ Belén Vaquerizo García*

Knowledge Based Expert System for PID Controller Tuning under
Hazardous Operating Conditions.................................... 203
    *Héctor Alaiz, José Luis Calvo, Javier Alfonso, Ángel Alonso, and*
    *Ramón Ferreiro*

**Author Index** ................................................... 211

# A Data Mining Based Analysis of Nmap Operating System Fingerprint Database

João Paulo S. Medeiros, Agostinho M. Brito Jr., and Paulo S. Motta Pires

LabSIN - Security Information Laboratory
Department of Computer Engineering and Automation – DCA
Federal University of Rio Grande do Norte – UFRN
Natal, 59.078-970, RN, Brazil
{joaomedeiros,ambj,pmotta}@dca.ufrn.br

**Abstract.** An Operating System (OS) fingerprint database is used by Nmap to identify OSes performing TCP/IP (Transmission Control Protocol/Internet Protocol) stack identification. Each entry in Nmap OS fingerprint database (*nmap-os-db*) represents an OS. Using data mining techniques, we propose three new forms of representation of *nmap-os-db* that can express how operating systems are similar among them according to their TCP/IP stack implementation. This approach can improve the capability of identifying devices running unknown OSes. Other applications are also presented.

## 1 Introduction

We use data mining techniques on the *nmap-os-db* to present how TCP/IP stack implementations are similar among them. New representations are obtained using three different techniques. The first representation consists in a rectangular lattice where each point represents one operating system. On this representation similar OSes will be topologically close. This is achieved using Kohonen's Self-Organizing Map (SOM) [1]. The Growing Neural Gas neural network [2] is used to produce a second representation. Using this neural network it's possible to build graphs where similar OSes will be connected in the same graph path. The last representation uses the classical $k$-means algorithm [3]. The $k$-means algorithm will provide an information of how operating systems can be grouped in clusters.

Our work is organized as follows. In Section 2 the OS fingerprinting concepts are presented and exemplified by Nmap OS detection system. Section 3 explains how Self-Organizing Map is used and presents the results obtained. The same is done for Growing Neural Gas neural network and its representation in Section 4. The $k$-means results are presented in Section 5. An overview of applications that can be done with our proposed representations is discussed in Section 6. Finally, conclusions are presented in Section 7.

## 2  OS Fingerprinting and Nmap

OS fingerprinting is the process of identifying the OS running on a target machine. One way to achieve this is using informations extracted from packets originated from the target machine over a network. When the network is based on TCP/IP protocols and the information used to distinguish OSes is extracted from messages of these protocols, we call this process TCP/IP stack fingerprinting [4]. The components and subprocess of OS fingerprinting are presented in Figure 1. In this figure components are numbered and the subprocess are delimited by dashed lines and identified by boldface names.

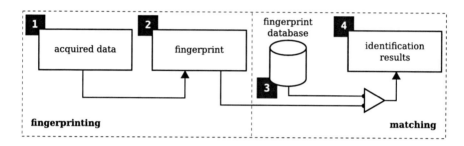

**Fig. 1.** Graphical representation of OS fingerprinting process

The overall process starts with the acquisition of target originated packets. OS fingerprinting methods can be classified in active and passive. In active methods, packets are sent to target machine aiming obtain desired responses. Passive methods work by listening to the network waiting for target messages. Nmap OS detection system uses the former method. The next step consists on building an representation for target OS using information contained in acquired packets. This representation is called fingerprint (component 2 of Figure 1), and in Nmap it is represented by a set of tests divided in classes as shown in Figure 2. The *nmap-os-db* consists of several of these entries.

```
1  Fingerprint OpenBSD 4.4
2  Class OpenBSD | OpenBSD | 4.X | general purpose
3  SEQ(R=N)
4  OPS(O1=%02=%03=%04=%05=%06=)
5  WIN(W1=0%W2=0%W3=0%W4=0%W5=0%W6=0)
6  ECN(R=Y%DF=Y%T=3B-45%TG=40%W=0%O=%CC=N%Q=)
7  T1(R=Y%DF=Y%T=3B-45%TG=40%S=Z%A=S+%F=AR%RD=0%Q=)
8  T2(R=N)
9  T3(R=N)
10 T4(R=Y%DF=Y%T=3B-45%TG=40%W=0%S=A%A=Z%F=R%O=%RD=0%Q=)
11 T5(R=Y%DF=Y%T=3B-45%TG=40%W=0%S=Z%A=S+%F=AR%O=%RD=0%Q=)
12 T6(R=Y%DF=Y%T=3B-45%TG=40%W=0%S=A%A=Z%F=R%O=%RD=0%Q=)
13 T7(R=N)
14 U1(DF=N%T=FA-104%TG=FF%IPL=38%UN=0%RIPL=G%RID=G%RIPCK=G%RUCK=G%RUD=G)
15 IE(DFI=S%T=FA-104%TG=FF%CD=S)
```

**Fig. 2.** A Nmap OS fingerprinter representation for OpenBSD 4.4

The second generation of Nmap OS detection via TCP/IP stack fingerprinting has one of the largest known fingerprint database [5]. The matching algorithm and the way Nmap shows its OS detection results is out of scope for this work. The focus of this work is on the component 3 of Figure 1.

To use *nmap-os-db* data in data mining algorithms, it is necessary to convert its representations to a numerical representation. The conversion was made using standard methods mapping numerical value to numerical range, class to numerical discretization or binary representation [6]. This was done using an vector of 402 length for each signature. To build representations, the Nmap OS fingerprint database entries that do not discriminate exactly an operating system was discarded. The original *nmap-os-db* has 430 entries. After the described discarding process a resulting database with 263 signatures was build.

## 3 Self-organizing Maps

Self-Organizing Maps have the capability of learning the intrinsics statistics of a given input space. This neural network is commonly used to express relations of high-dimensional data using bi-dimensional graphical representations [7]. The structure of SOM neural network is composed by one input layer and one output layer. The input layer is a vector with dimension $m$, where $m$ is the dimensionality of the input space. The output layer is composed by neurons. Each neuron has a set of synapses that connects to each position of the vector of the input layer, as presented in Figure 3.

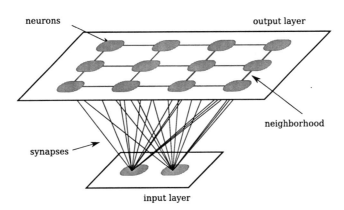

**Fig. 3.** SOM neural network structure

The standard SOM training algorithm was used [8]. After 10000 iterations for a 10 × 10 lattice the algorithm produces the map shown in Figure 4. After the training phase each neuron was labeled with the most close Nmap fingerprint.

In Figure 4, we can see that signatures of the same operating system stay concentrated together. The BSD based systems are placed at top right side of map. The different Windows versions appear at bottom left, while the Cisco

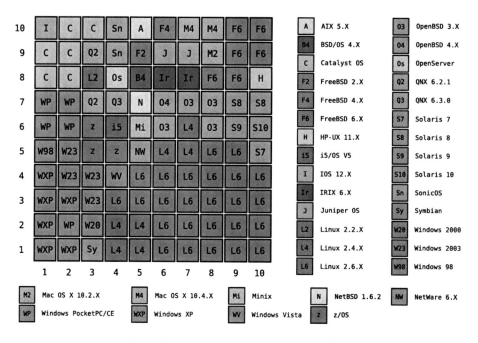

**Fig. 4.** Database representation using SOM

operating systems appear at top left. We can see that there is a relation between Mac OS X, FreeBSD and JunOS. This confirm a known fact that Mac OS X [9], and JunOS [10] are based on FreeBSD. It is also visible that new Linux versions (i.e. 2.4 and 2.6 series) seems significantly different of Linux 2.2 series. Also, we can see that NetBSD is near to QNX. This emphasize the fact that QNX adopted NetBSD network stack implementation [11].

## 4 Growing Neural Gas

While SOM neural network uses a rectangular lattice to express similarities, the Growing Neural Gas (GNG) neural network can be used to build graphs for the same purpose. GNG was proposed by [2] as a solution to the fixed number of neurons of Neural Gas algorithm [12]. Using this approach we intend to see how operating systems can be clustered and has a better measure of how much close they are.

Figure 5 shows the GNG results using the standard training algorithm [2] after 100 iterations with parameter $\lambda = 100$ and the max number of nodes 100 (the same number of neurons used in SOM map). When the training algorithm is finished each neuron was labeled with the name of the closest OS fingerprint.

Figure 5 introduces some new informations not shown by SOM. For example, the graph shows that Mac OS X 10.2 is derived from FreeBSD 5, that is really true [9]. Mac OS X 10.4 seems derived from FreeBSD 6 and Catalyst OS has a

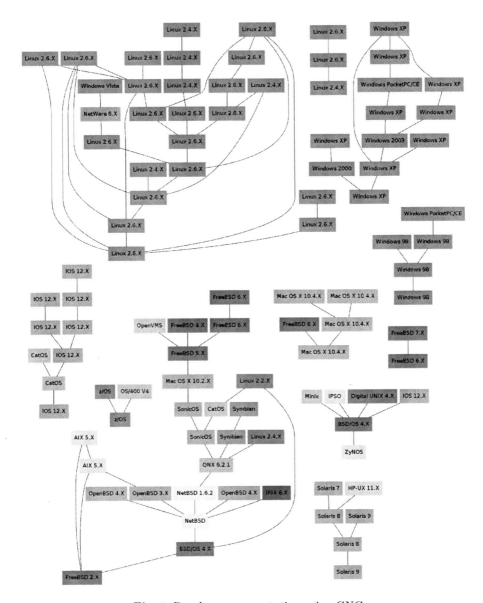

**Fig. 5.** Database representation using GNG

close relation to Linux 2.2. The QNX and NetBSD similarity is also enforced in this graph.

## 5 K-Means

The $k$-means algorithm find the $k$ centroids for a given input data set. These centroids can be used to represent clusters, since each entry in input data set can be associated to the closest centroid. We use $k$-means because it presents all the input into results [3]. The number of clusters $k$ was chosen based on the information given by GNG results. As shown in Figure 5, there are 11 graphs in GNG results, because of this the number of clusters $k$ was also 11.

**Table 1.** Operating systems for each cluster of $k$-means result

| N" | Size | Operating Systems | N" | Size | Operating Systems |
|---|---|---|---|---|---|
| 1 | 31 | AIX 4.X e 5.X; AsyncOS; BSD/OS 4.X; CatOS; | 4 | 65 | Linux 2.4.X, 2.6.X e 2.6.16; NetWare 5.X e 6.X; RouterOS; Windows Vista. |
| | | Digital UNIX 4.X; FreeBSD 2.X e 3.X; IRIX 6.X; | 5 | 21 | CatOS; IOS 12.X; ZyNOS. |
| | | Linux 2.1.X e 2.2.X; | 6 | 4 | Mac OS X 10.3.X e 10.4.X. |
| | | NetBSD 1.6.2; NmpSW; OpenBSD 2.X, 3.X e 4.X; OpenServer; QNX 6.2.1 e 6.3.0; SonicOS; pSOS. | 7 | 11 | Linux 2.4.X e 2.6.X; Minix; OS/400 V4; Windows 95 e 98; i5/OS V5; z/OS; z/VM. |
| 2 | 17 | HP-UX 11.X; OpenBSD 4.X; | 8 | 34 | Linux 2.4.X e 2.6.X; OpenBSD 4.1. |
| | | Solaris 2.X, 7, 8, 9, 10. | 9 | 40 | Symbian; |
| 3 | 32 | FreeBSD 4.X, 5.X, 6.X e 7.X; | | | Windows 98, NT, 2000, 2003, XP e CE. |
| | | HP-UX 11.X; JunOS; Mac OS X 10.2.X, 10.4.X e 10.5.X; OpenBSD 3.X e 4.X; OpenVMS; | 10 | 6 | IPSO; Linux 2.6.X; Solaris 9; ZyNOS; z/OS. |
| | | Symbian; True64 UNIX 5.X. | 11 | 2 | Linux 2.6.X. |

Table 1 shows the cluster number, the number of nodes in each cluster (size), and the operating systems that each cluster contain. Using this approach informations not seem on previous results can be analysed. For example, the AsyncOS that is based on FreeBSD [10] did not appeared on previous results, but now it is at the same cluster of a FreeBSD 2 and 3.

# 6 Applications

Beyond all tasks in which OS fingerprinting processes can be applied to, we propose a few more applications for the methods described here. First, we have found that the SOM map can be used to address security tests for unknown devices [13,14]. This can be accomplished by the generalization capabilities of SOM, since one of the features of the map is create an approximation of the input space. For example, consider, in the fingerprint map shown in Figure 6, the localization of two devices with unknown operating systems. This figure shows the operating system's neighborhood of these two devices extracted from a 20 × 20 version of SOM map. As Nmap can't show an exact result about these target devices, one solution is to assume that these OSes can be represented by the closest neuron of the approximated input space or, in a general sense, by the $(2r+1)^2$ neighborhood neurons, including the closest one, where $r$ is the radius starting from the closest neuron.

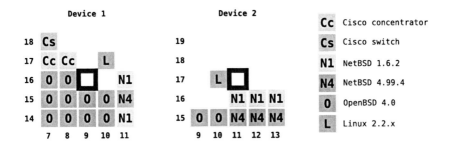

**Fig. 6.** Fingerprint placement of two proprietary devices with $r = 2$ (the blank placements are related to fingerprints that represents devices with unknown operating system, e.g. printers, cameras and so on) [14]

The input space approximation feature of SOM map can be used also to reduce the original size of fingerprinting database, since the most representative fingerprints are those that are closest of SOM neurons from output layer [7]. As shown in Section 3, we are able to reduce the input space composed of 263 fingerprints by 100 (approximately 38%). This works like a compression algorithm that retains the most important information of the *nmap-os-db* in a statistical sense. The reduction of OS fingerprint databases is a expressive improvement to critical applications that perform OS fingerprinting mainly when their are subject to real-time constraints.

The results presented by GNG algorithm can be used to measure the quality of Nmap OS fingerprint database. Figure 5 shows a graph with three Linux nodes that are apart of main Linux graph. This also can be found at cluster 11 of Table 1. Probably these signatures are invalid ones. There are huge number of network functionalities that can corrupt Nmap OS fingerprinting process, e.g., traffic normalization, TCP SYN proxies, and Port Address Translation (PAT) [15].

## 7 Conclusions

Three data mining methods were used to build different representations of Nmap OS fingerprint database. Three main applications of these new representations are proposed. First, to improve the identification of unknown operating systems.

Second, to reduce Nmap OS fingerprinting database by compressing its information. What can improve the efficiency of some real-time applications. And, finally, identify possible corrupt OS fingerprint entries into *nmap-os-db*.

**Acknowledgments.** The authors would like to express their gratitude to the Department of Computer Engineering and Automation, Federal University of Rio Grande do Norte, Brazil, and REDIC (Instrumentation and Control Research Network) for supporting this work.

# References

1. Kohonen, T.: Self-organized formation of topologically correct feature maps. Biological Cybernetics 43(1), 59–69 (1982)
2. Fritzke, B.: A Growing Neural Gas Network Learns Topologies. Advances in Neural Information Processing Systems 7 (1995)
3. Hartigan, J., Wong, M.: A K-means Clustering Algorithm. JR Stat. Soc. Ser. C-Appl. Stat. 28, 100–108 (1979)
4. Fyodor: Remote OS Detection via TCP/IP Fingerprinting. Phrack Magazine 8 (1998)
5. Fyodor: Nmap Network Scanning. Insecure.Com LLC (2008)
6. Han, J., Kamber, M.: Data mining: concepts and techniques, 2nd edn. Morgan Kaufmann, San Francisco (2006)
7. Kohonen, T.: Self-Organizing Maps, 3rd edn. Springer, Heidelberg (2001)
8. Haykin, S.: Neural Networks: A Comprehensive Foundation. Prentice-Hall, Englewood Cliffs (1999)
9. Apple Developer Connection: Open Source (2009), http://developer.apple.com/opensource/index.html
10. FreeBSD News: FreeBSD embedded systems (2008), http://www.freebsdnews.net/2008/07/24/freebsd-embedded-systems/
11. NetBSD Project: Products based on NetBSD (2009), http://www.netbsd.org/gallery/products.html
12. Martinetz, T., Schulten, K.: A Neural-Gas Network Learns Topologies. Artificial Neural Networks 1, 397–402 (1991)
13. Medeiros, J.P.S., Cunha, A.C., Brito, A.M., Pires, P.S.M.: Automating Security Tests for Industrial Automation Devices Using Neural Networks. In: Proc. IEEE Conference on Emerging Technologies & Factory Automation, pp. 772–775 (2007)
14. Medeiros, J.P.S., Cunha, A.C., Brito, A.M., Pires, P.S.M.: Application of Kohonen Maps to Improve Security Tests on Automation Devices. In: Lopez, J., Hämmerli, B.M. (eds.) CRITIS 2007. LNCS, vol. 5141. Springer, Heidelberg (2008)
15. OpenBSD PF: The OpenBSD Packet Filter – OpenBSD 4.4 (2008)

# Knowledge System for Application of Computer Security Rules

Menchaca García Felipe Rolando and Contreras Hernández Salvador

Computer research Center-IPN, Av. Juan de Dios Bátiz, esquina con Miguel Otón de Mendizábal, México, D.F., 07738. México
menchaca@cic.ipn.mx, scontreras@upvm.edu.mx

**Abstract.** Epistemic Logic is applied to diverse fields of the computing science, particularly to those related to systems constituted by agents. Such systems have an interesting range of applications themselves. In this paper, a logic system applied to the evaluation of Multilevel and Commercial computer security rules is discussed. The system is defined by axioms and rules, and the concepts of the security models are then considered in order to evaluate the rules. It is showed how to lead to certain formulas in order to determine the validity of the set of fundamental expressions that constitute the security models. It uses a system S5.

**Keywords:** possible worlds, knowledge, common knowledge, ontology, security rules.

## 1 Introduction

Nowadays, teamwork using communication networks, has become an every day matter. Access to every single resource of an organization's web depends on its security rules' group, which in a common way permits the Access to public or non confidential resources to any user identified as a guest, as in internet systems. Another way to permit remote access to a computer system is by giving a user account within the system to people located in places which are geographically far from where the servers are. This is a very basic way of interaction between two or more organizations and it is present when a user that does not belong to the system requires certain specific resources. It is necessary to make it clear that these two schemes do not represent a way of having users of two security domains access resources belonging to them both, respecting a set of rules in each one, they are mechanisms used to solve in any way, the access to resources problems. We can make that not only users interact with different domains to the ones they belong to, but processes can do it as well, it is to say, the latter can access resources. To formalize interaction concepts between subjects and objects from different security domains, this work proposes a system of knowledge logic that models access to resources in computer systems with a set of different security rules. Two well-known security models are used; Multilevel and Commercial. To move from theory to practice, an ontology that contemplates a set of concepts of the security models that can be taken as a standard must be created, in this way organizations which use them can, independently from their security policies, establish interaction between their subjects and objects. In this way, concepts presented here may be applied to systems of common purpose that need sharing resources, for instance, systems of government groups, medical associations, lawyers, colleges, etc.

The problem of security among domains with different sets of rules has been boarded by ECMA (European Computer Manufacturers Association) that defines a security frame in distributed systems in which bases for an existing cooperation among different security domains are presented.

Work related to this one, the formal approach for the reasoning of security systems, is Security Logic (SL), which was proposed by Glassgow, McEwen and Panangaden [7]. Fundamental elements are: knowledge, permission and obligation. With these elements it is possible to model security in its integrity and confidentiality aspects. Another important point of this approach is that it permits to combine security policies. It uses the notion of *possible worlds* to define knowledge logic [8].

In order to model the concept in terms of epistemic logic, we will use an S5 system all together with the concepts of knowledge K, common knowledge C and set knowledge E. The language for the logic of knowledge S5 for $n$ agents contains the next elements:

a) A set of propositions p, q, r,...
b) $\land, \lor, \rightarrow, \leftrightarrow, \neg$ connectives of propositional logic
c) $\varphi, \lambda, \mu,...$ formulas or metalinguistic variables
d) K, C, E epistemic operators as defined in [6]

The right combination of language symbols lead to well-formed formulas [4] which will be of any of the following types:

1. Every variable p, q, r.
2. $\neg \varphi$ where $\varphi$ is a well-formed formula
3. $(\varphi \land \lambda), (\varphi \lor \lambda), (\varphi \rightarrow \lambda), (\varphi \leftrightarrow \lambda), (\varphi \neg \lambda)$
4. $K_i \varphi$ agent i knows $\varphi$, where $\varphi$ is a well-formed formula.
5. $C \varphi$  $\varphi$ is common knowledge
6. $E \varphi$  everybody knows $\varphi$

**Definition 1.** May P a non-empty set of propositional variables and $m \in \mathbf{N}$. The Language L is the smallest super set of P so as $\varphi, \psi \in L \rightarrow \neg \varphi, (\varphi \land \psi), K_i \varphi, C \varphi, E \varphi \in L$ (i≤m).

We will use the set of operators OP={$K_1,...K_m$, C,E} as modal operators. Expressing $K_i \varphi$ as the fact that agent i knows that $\varphi$, $E \varphi$ means that everybody knows $\varphi$, $C \varphi$ is common knowledge among a group of agents.

**Table 1.** Axioms for L

| A1. Propositional logic axiom | A6. $E\varphi \leftrightarrow (K_1\varphi \wedge ... \wedge K_m\varphi)$ |
|---|---|
| A2. $(K_i\varphi \wedge K_i(\varphi \rightarrow \lambda)) \rightarrow K_i\lambda$ | A7. $C\varphi \rightarrow \neg E_t\varphi$ |
| A3. $K_i\varphi \rightarrow \varphi$ | A8. $C\varphi \rightarrow EC\varphi$ |
| A4. $K_i\varphi \rightarrow K_iK_i\varphi$ | A9. $(C\varphi \wedge C(\varphi \rightarrow \lambda)) \rightarrow C\lambda$ |
| A5. $\neg K_i\varphi \rightarrow K_i\neg K_i\varphi$ | A10. $C(\varphi \rightarrow E\varphi) \rightarrow C(\varphi \rightarrow C\varphi)$ |

Language allows agents to express their knowledge in a security rules system, which fulfillment is responsibility of the agents themselves; also, it allows to expressing what an agent knows about his own knowledge and about other agents' knowledge.

## 2 Models for L

The approach proposed in this paper is based on the sharing of an ontology between the domains that operate on the process of interaction. For this, we have considered the set of security rules "multilevel" (MLS), so called "military" [3], which is well-known, as well as the Commercial [1]. In the case of a domain to enforce multilevel rules, when this receives a request to access a resource, a verification of the applicant's attributes should be made. As part of such verification, it has to be determined the set of rules to which the subject is subjugated. If the policy of the individual requesting resources is multilevel, then access to the objects is feasible as long as their category and classification allow it. Of course, such attributes have to be known to the environment that receives the access request. Despite this article focuses on MLS and commercial, the approach to rules of evaluation applies to other groups such as Financial, Take Grant, etc. In a domain with a commercial security policy, which receives a request to access resources from another environment where the rules may be similar or completely different, the ontology can be applied along with the set of Commercial rules (such as a layer that allows interaction), because, even though the policies are shared, the configuration will be different. To do this, the domain in charge of granting access or denying resources, will have to verify the TPs or transforming procedures, the CDIs or restricted data sets, and the UDIs or non-restricted data sets.

The existence of a common ontology on both ends of the interaction is not trivial, it is a key element that permits agents to communicate by means of proper rules. This model assumes the following rules of derivation [5]:

R1.- $\vdash \varphi, \vdash \varphi \rightarrow \psi \Rightarrow \vdash \psi$
R2.- $\vdash \varphi \Rightarrow \vdash K_i\varphi$, for all $i <= m$ where m is the total number of agents
R3.- $\vdash \varphi \Rightarrow \vdash C\varphi$

We will call L to the logical language for knowledge systems. The next expressions are derived in L ($\alpha, \beta \in$ L y $\square \in$ OP).

$\vdash \alpha \Rightarrow \vdash \square\alpha$
$\vdash \alpha \rightarrow \beta \Rightarrow \vdash \square\alpha \rightarrow \square\beta$
$\vdash \alpha \rightarrow \beta \Rightarrow \vdash \neg\square\neg\alpha \rightarrow \neg\square\neg\beta$
$\vdash \alpha \rightarrow \gamma, \vdash \gamma \rightarrow \beta \Rightarrow \vdash \alpha \rightarrow \beta$
$\vdash K_i \neg K_i \neg K_i\alpha \rightarrow K_i\alpha$
$\vdash C\alpha \rightarrow E\alpha$

The set of knowledge needed to verify the security rules among domains are contained in Table 1.

**Table 2.** Knowledge – set of multilevel security rules

| i. Access level | x. Security rule |
|---|---|
| ii. Category | xi. Set of security rules |
| iii. Subject | xii. Evaluation |
| iv. Object | xiii. Access to reading verification |
| v. Access to reading | xiv. Access to writing verification |
| vi. Access to writing | xv. Obtaining security level |
| vii. Access to Reading-writing | xvi. Obtaining security category |
| viii. Domain | xvii. Assigning attributes |
| ix. Security policy | |

In the expressions used in the logic model, a combination of some of these concepts - represented by Greek letters with other simple ones (such as subjects, objects and access to reading/writing) - is made. More specifically, the formulas and their meanings are used as follows:

$\gamma(s)$. Access level of subject $s$.
$\varepsilon(s)$. Security category of subject $s$.
$\alpha(s,o)$. Access to the reading of subject $s$ on object $o$.
$\beta(s,o)$. Access to the writing of subject $s$ on object $o$.
$\delta(s,o)$. Access to Reading-writing of subject $s$ on object $o$.

Literals in Arabic alphabet have been used to identify the concept it represents, this is, for the formulas of access to reading, writing, reading-writing and for the security level and category, respectively. All of them represent concepts in the ontology and formulas in the formal representation. In the verification of the security attributes within the multi domain environment, using MLS as a set of rules, the following key expressions represent the agents' valid actions in a system of verification of security attributes for a multilevel model.

1.- $\vdash \alpha(s,o) \rightarrow \vdash \square\alpha(s,o)$
2.- $\vdash (\gamma(s)>\gamma(o)) \wedge (\varepsilon(s)>\varepsilon(o)) \rightarrow \alpha(s,o) \Rightarrow \vdash \square(\gamma(s)>\gamma(o)) \wedge (\varepsilon(s)>\varepsilon(o)) \rightarrow \square\alpha(s,o)$

3.- $\vdash (\gamma(s)<\gamma(o))\wedge(\varepsilon(s)<\varepsilon(o))\rightarrow\beta(s,o) \Rightarrow \vdash \Box(\gamma(s)<\gamma(o))\wedge(\varepsilon(s)<\varepsilon(o))\rightarrow\Box\beta(s,o)$
4.- $\vdash (\gamma(s)=\gamma(o))\wedge(\varepsilon(s)=\varepsilon(o))\rightarrow\delta(s,o) \Rightarrow \vdash \Box(\gamma(s) = \gamma(o))\wedge(\varepsilon(s)=\varepsilon(o)) \rightarrow \Box\delta(s,o)$
5.- $\vdash \Box(\gamma(s) > \gamma(o))\wedge(\varepsilon(s)>\varepsilon(o))\rightarrow \Box\alpha(s,o) \Rightarrow \vdash \neg\Box\neg(\gamma(s) > \gamma(o))\wedge(\varepsilon(s)>\varepsilon(o))\rightarrow \neg\Box\neg\alpha(s,o)$
6.- $\vdash (\gamma(s)=\gamma(o))\wedge(\varepsilon(s)=\varepsilon(o))\rightarrow\delta(s,o), \vdash \delta(s,o)\rightarrow\beta(s,o), \vdash (\gamma(s) = \gamma(o))\wedge(\varepsilon(s)=\varepsilon(o))\rightarrow \beta(s,o)$
7.- $\vdash (\gamma(s) = \gamma(o))\wedge(\varepsilon(s)=\varepsilon(o))\rightarrow\delta(s,o), \vdash \delta(s,o)\rightarrow\alpha(s,o), \vdash (\gamma(s)=\gamma(o))\wedge(\varepsilon(s)=\varepsilon(o))\rightarrow \alpha(s,o)$
8.- $\vdash K_i\neg K_i\neg K_i \alpha(s,o) \rightarrow K_i \alpha(s,o)$
9.- $C\alpha(s,o)\rightarrow E\alpha(s,o)$
10.- $C\beta(s,o)\rightarrow E\beta(s,o)$

Now we will analyze some of the derived expressions previously shown, which were obtained by applying the axioms and the derivative rules.

2.- $\vdash \alpha \rightarrow \beta \Rightarrow \vdash \Box\alpha \rightarrow \Box\beta$
According to R3, we can derive $\vdash \Box(\alpha\rightarrow\beta)$ from $\vdash \alpha\rightarrow\beta$. Also, $\vdash \Box(\alpha\rightarrow\beta) \rightarrow (\Box\alpha\rightarrow\Box\beta)$ and the conclusion follows from R1. Substituting $\alpha$ and $\beta$ for the expressions shown to access the reading material, it remains as $\vdash (\gamma(s) > \gamma(o))\wedge(\varepsilon(s)>\varepsilon(o)) \rightarrow \alpha(s,o)$ $\vdash \Box(\gamma(s) > \gamma(o))\wedge(\varepsilon(s)>\varepsilon(o)) \rightarrow \Box\alpha(s,o)$. The same goes for expressions 3 and 4.

5.- By using A1 and R1 we can obtain $\vdash \neg\alpha \rightarrow \neg\beta$ from $\vdash \alpha \rightarrow \beta$, then, by applying A1 once again, we obtain $\vdash \neg\Box\neg\alpha \rightarrow \neg\Box\neg\beta$. By using the fundamental expression 5, we conclude that $\vdash \Box(\gamma(s)>\gamma(o))\wedge(\varepsilon(s)>\varepsilon(o))\rightarrow \Box\alpha(s,o) \Rightarrow \vdash \neg\Box(\gamma(s) > \gamma(o))\wedge(\varepsilon(s)>\varepsilon(o))\rightarrow \neg\Box\alpha(s,o)$ and from this we can derive $\vdash \neg\Box\neg(\gamma(s)>\gamma(o))\wedge(\varepsilon(s)>\varepsilon(o))\rightarrow \neg\Box\neg\alpha(s,o)$

8.- Expressions of type 8 belong to the previously mentioned $\vdash K_i\neg K_i\neg K_i\alpha \rightarrow K_i\alpha$ type. To demonstrate this we use the principle stating that $\vdash (\varphi \wedge \neg\psi) \rightarrow \bot$ implies that $\vdash \varphi \rightarrow \psi$. This can be validated with A1 and R1. Therefore, it is enough to prove that $\vdash K_i\neg K_i\neg K_i\alpha \wedge \neg K_i\alpha$. For this, we use again the access to reading formula $\alpha(s,o)$.

$\vdash K_i\neg K_i\neg K_i\alpha(s,o) \wedge \neg K_i\alpha(s,o) \rightarrow (\neg K_i\neg K_i\alpha(s,o) \wedge \neg K_i\alpha(s,o))$     A1,R1,R3
$\vdash (\neg K_i\neg K_i\alpha(s,o) \wedge \neg K_i\alpha(s,o)) \rightarrow (\neg K_i\neg K_i\alpha(s,o) \wedge K_i \neg K_i\alpha(s,o))$     A1,R1,A5
$\vdash (\neg K_i\neg K_i\alpha(s,o)\wedge K_i \neg K_i\alpha(s,o)) \rightarrow \bot$     A1

Now, two important constructions in L are presented as well as the necessary derivations to get to them from the axioms and derivation rules.

**Proposition 1.** In L, the common knowledge C satisfies the negative and positive introspections.

$\vdash C\varphi \rightarrow CC\varphi$
$\vdash \neg C\varphi \rightarrow C\neg C\varphi$

The test for the positive introspection, using the access to reading formula is shown below.

$\vdash C(C\alpha(s,o) \rightarrow EC\alpha(s,o)) \rightarrow (C\alpha(s,o) \rightarrow CC\alpha(s,o))$
$\vdash C(C\alpha(s,o) \rightarrow EC\alpha(s,o))$
$(C\alpha(s,o) \rightarrow CC\alpha(s,o))$

Demonstration of formulas 3, 4, 6, 7 and 9 is made in a similar way to the previous ones. On the other hand, by using axioms and derivation rules, we can prove the negative introspection for C in the control rule of access to reading. The negative introspection corresponds to axiom 5 which identifies the S5 system.

1. $\vdash \neg K_i \neg C\alpha(s,o) \rightarrow K_i \neg K_i \neg C\alpha(s,o)$
2. $\vdash C\alpha(s,o) \rightarrow K_i C\alpha(s,o)$
3. $\vdash \neg K_i \neg C\alpha(s,o) \rightarrow K_i \neg K_i C\alpha(s,o)$
4. $\vdash K_i \neg K_i \neg C\alpha(s,o) \rightarrow K_i \neg K_i \neg K_i C\alpha(s,o)$
5. $\vdash \neg K_i \neg C\alpha(s,o) \rightarrow K_i \neg K_i \neg K_i C\alpha(s,o)$
6. $\vdash K_i \neg K_i \neg K_i C\alpha(s,o) \rightarrow K_i C\alpha(s,o)$
7. $\vdash \neg K_i \neg C\alpha(s,o) \rightarrow K_i C\alpha(s,o)$
8. $\vdash \neg K_i \neg C\alpha(s,o)$
9. $\vdash \neg C\alpha(s,o) \rightarrow K_i \neg C\alpha(s,o)$
10. $\vdash C(\neg C\alpha(s,o) \rightarrow K_i \neg C\alpha(s,o))$
11. $\vdash C(\neg C\alpha(s,o) \rightarrow K_i \neg C\alpha(s,o)) \rightarrow (\neg C\alpha(s,o) \rightarrow C(\neg C\alpha(s,o))$
12. $\vdash \neg C\alpha(s,o) \rightarrow C \neg C\alpha(s,o)$

Constructions that represent the rules of a Commercial model are shown in the following lines from the point of view of the approach to evaluate rules in heterogeneous domains. For doing so, it is important to remember that a subject in this model does not have direct access to a resource, but uses a transformation procedure (TP) that has permission to manipulate it. These resources are represented as a set of restricted data CDIs. Not sensible data or resources are called UDIs. We will use the following ontology elements of the Commercial model.

$\alpha(s_k, cdi_j)$. Access to reading of the subject $s$ on the object $o$.
$\beta(s_k, cdi_j)$. Access to reading of the subject $s$ on the object $o$.
$\delta(s_k, cdi_j)$. Access to reading-writing of the subject $s$ on the object $o$.
T. Set of TPs of the subject s.
$\Lambda$. Set of restricted data.
$tp_i$. i-th transformation procedure $y \in T$.
$cdi_j$. j-th set of restricted data $y \in \Lambda$.
$s_k$. k-th subject.
$s_k(tp_i)$. Subject i has the transformation procedure tpi associated.
$tp_i(cdi_j)$. The $cdi_j$ is asigned to the list of transformation procedures tpi.
1.- $\vdash \alpha(s_k, cdi_j) \rightarrow \vdash \Box\alpha(s_k, cdi_j)$
2.- $\vdash (s_k(tp_i) \land tp_i(cdi_j)) \rightarrow \alpha(s_k, cdi_j) \Rightarrow \vdash \Box((s_k(tp_i) \land tpi(cdi_j)) \rightarrow \Box\alpha(s_k, cdi_j)$

3.- $\vdash K_i \neg K_i \neg K_i\ \alpha(s_k,cdi_j) \to K_i\ \alpha(s_k,cdi_j)$
4.- $C\alpha(s_k,cdi_j) \to E\alpha(s_k,cdi_j)$
5.- $C\beta(s_k,cdi_j) \to E\beta(s_k,cdi_j)$

Access to reading, writing or both, depends on the transformation procedures associated to the subject. Here is only shown the access to reading in expression 2, considering all formulas are the same. The expressions 3, 4 and 5 are in terms of knowledge, this, represented by the modal operators K, C and E, which indicate knowledge, common knowledge and knowledge of all the agents in the context, respectively. Despite the fact that three types of access to resources have been considered in the Commercial model (reading, writing and reading-writing), does not mean that there is a relation between this and the Multilevel model. As there are, in fact, some points that may coincide, the approach does not consider matching attributes of two or more security policies. Nevertheless, it is correct to build an ontology taking into account two or more sets of security rules. Derived expressions seen previously are equally valid for the Commercial model.

$\vdash \alpha(s_k,cdi_j) \to\ \vdash \Box\alpha(s_k,cdi_j)$
$\vdash \alpha \to \beta \Rightarrow\ \vdash \Box\alpha \to \Box\beta$

Based on R3, of $\vdash \alpha \to \beta$ we can derive $\vdash \Box(\alpha \to \beta)$. Also $\vdash \Box(\alpha \to \beta) \to (\Box\alpha \to \Box\beta)$ and the conclusion follows from R1. Substituting $\alpha$ and $\beta$ for the expressions shown for the access to reading in the Commercial policy, it would be

$\vdash (s_k(tp_i) \wedge tp_i(cdi_j)) \to \alpha(s_k,cdi_j)\ \vdash \Box((s_k(tp_i) \wedge tpi(cdi_j)) \to\ \Box\alpha(s_k,cdi_j).$
$\vdash K_i \neg K_i \neg K_i\ \alpha(s_k,cdi_j) \to K_i\ \alpha(s_k,cdi_j)$

This expression is of the type $\vdash K_i \neg K_i \neg K_i \alpha \to K_i \alpha$ aboarded in the set of MLS rules mentioned before. The demonstration is similar, using the principle which says $\vdash (\varphi \wedge \neg \psi) \to \bot$ implies $\vdash \varphi \to \psi$. This may be validated with A1 and R1. Therefore, it is enough to prove that $\vdash K_i \neg K_i \neg K_i \alpha(s_k,cdi_j) \wedge \neg K_i \alpha(s_k,cdi_j)$. Using once more, for this purpose, the access to reading formula $\alpha(s_k,cdi_j)$.

$\vdash K_i \neg K_i \neg K_i \alpha(s_k,cdi_j) \wedge \neg K_i \alpha(s_k,cdi_j) \to (\neg K_i \neg K_i \alpha(s_k,cdi_j) \wedge \neg K_i \alpha(s_k,cdi_j))$ A1,R1,R3
$\vdash (\neg K_i \neg K_i \alpha((s_k,cdi_j) \wedge \neg K_i \alpha(s_k,cdi_j)) \to (\neg K_i \neg K_i \alpha(s_k,cdi_j) \wedge K_i \neg K_i \alpha(s_k,cdi_j))$ A1,R1,A5
$\vdash (\neg K_i \neg K_i \alpha(s_k,cdi_j) \wedge K_i\ \neg K_i \alpha(s_k,cdi_j)) \to \bot$ A1
$C\alpha(s_k,cdi_j) \to E\alpha(s_k,cdi_j)$
$C\beta(s_k,cdi_j) \to E\beta(s_k,cdi_j)$

The integration of ontologies in this approach does not need to match security attributes, but it consists in counting with two or more ontologies of computer security models. If these ontologies may as well be independent and dwell in one domain, an ontology containing rules of two or more security policies may be developed [2], but communication between two or more domains should use only one of them in a certain moment. To have an integration of ontologies (something which is not strictly necessary to perform the evaluation of rules in multiple domains) new rules are simply added to the existent ones, for instance, in a Multilevel model it would be enough to add the rules of Commercial policy. Thus, it is possible to have a single ontology that permits access to resources from other domains that contain any of the mentioned models, Multilevel or Commercial, or even any other such as Take Grant or Financial. The rules and axioms stablished for the representation of knowledge and evaluation of MLS and Commercial rules, are similar. Differences are found in the access to resources formulas, which without any doubt, makes these two models totally different. Related to these points, it is important to highlight that the ontology must be separated from the mechanism that applies the access rules. This carries some disadvantages, mainly, the modification of the ontology, updating elements or adding some others. The system could get new formulas, which may update the set of knowledge [6], even as this topic is beyond reach of the present project.

## 3 Validity of the Formulas in the Knowledge System

Validity of the formulas of the model presented may be proved using the semantic Tableaux methods. Tableaux's systems are used in other logics, they are not exclusive for S5, for example, S5 predecessor logics, such as; K, K4, T and S4. A Tableaux is a tree with its root in the upper part. A proof by this method is a tree containing a X formula in the root, labeled $\neg X$. From the root the tree grows downwards following the extension of branches rules. One of the Tableaux branches is closed if it contains a syntactic contradiction, like W y $\neg W$ for any W formula. If all the branches of the tree are closed, then we will say that the Tableaux is closed. A closed Tableaux where the root contains a formula $\neg X$, is a demonstration of the validity of X. The extension of branches rules for the Tableaux are composed by the conjunctive, disjunctive, necessary and possible rules.

## 4 Conclusions

Formalization of rules of access control to resources in multiple domain surroundings has permited to describe, from an epistemic point of view, the interactions that occur taking into account the scenery variety. Two sets of rules, Multilevel and Commercial have been presented in this project, and their fundamental aspects have been considered to create a knowledge base that permits to determine if a subject can take a resource from a specific domain. The model created to validate rules of security contains the necessary elements to represent the problem and the set of concepts which integrate it (S5 system). Results, from a logical point of view, let us know that the system is consistent and complete. It has been demonstrated that formulas built for the concepts of the sets of rules presented, are valid in the system. This suggests that implementation of a system of such nature may work and offer very interesting results in its use.

## References

1. Vázquez-Gómez, J.: Modelling Multidomain Security SIGSAC. In: New Security Paradigm workshop. IEEE Computer Society Press, Los Alamitos (1993)
2. Vázquez-Gómez, J.: Multidomain Security. Computers & Security 13, 161–184 (1994)

3. Bell, D.E., La Padula, L.J.: Security Computer Systems: Mathematical Foundations and Model. Tech. Rep., MITRE Corp., Bedford MA (1974)
4. Chellas, B.F.: Modal Logic an Introduction. Cambridge University Press, Cambridge (1995)
5. Baral, C., Zhang, Y.: On the semantics of knowledge update. In: Proceedings of the 17th International Joint Conference on Artificial Intelligence (IJCAI 2001), Seattle, WA, pp. 97–102. Morgan Kaufmann, San Mateo (2001)
6. Halpern, J.Y., Moses, Y.: Knowledge and common knowledge in a distributed environment. J. ACM 37(3), 549–587 (1990), http://doi.acm.org/10.1145/79147.79161
7. Glassgow, J., MacEwen, G., Panangaden, P.: A logic for reasoning about security. ACM Trans. Comput. Syst. 10(3) (August 1992)
8. Glassgow, J., Mac Ewen, G.: Obligation as the basis of integrity especification. In: Proc. Computer Security Fundations Workshop, Franconia, NH (June 1989)

# Clustering of Windows Security Events by Means of Frequent Pattern Mining

Rosa Basagoiti[1], Urko Zurutuza[1], Asier Aztiria[1],
Guzmán Santafé[2], and Mario Reyes[2]

[1] Mondragon University, Mondragon, Spain
{rbasagoiti,uzurutuza,aaztiria}@eps.mondragon.edu
[2] Grupo S21sec Gestión S.A., Orcoyen, Spain
{gsantafe,mreyes}@s21sec.com

**Abstract.** This paper summarizes the results obtained from the application of Data Mining techniques in order to detect usual behaviors in the use of computers. For that, based on real security event logs, two different clustering strategies have been developed. On the one hand, a clustering process has been carried out taking into account the characteristics that define the events in a quantitative way. On the other hand, an approach based on qualitative aspects has been developed, mainly based on the interruptions among security events. Both approaches have shown to be effective and complementary in order to cluster security audit trails of Windows systems and extract useful behavior patterns.

**Keywords:** Windows security event analysis, data mining, frequent pattern mining, intrusion detection, anomaly detection.

## 1 Introduction

The idea of discovering behavioral patterns from a set of event logs in order to detect unusual behavior or malicious events is not novel. In fact, the idea came up in the 80s when James P. Anderson, in a seminal work in the area of Intrusion Detection Systems [1], suggested that the common behavior of a user could be portrayed analyzing the set of event logs generated during his/her use of computer. Thereby, unusual events, out of such 'usual' behavior could be considered as attacks or at least as unusual. There are many works in this sense, but most of them have been developed considering Unix systems. This paper focuses on events produced by Windows operative systems. The complexity of such systems is even bigger due to the large amount of data they usually generate. In this work, different experiments have been carried out considering two different approaches. On the one hand, we have created clusters based on characteristics which summary the activity from a quantitative point of view. The second approach tries to find out logical clusters analyzing the interruptions among events.

The reminder of this paper is organized as follows. Section 2 provides a literature review of different tools and approaches when performing the analysis of log data. In Section 3 we analyse the nature of the problem and we define some aspects to be considered. Section 4 describes the experiments and the results we have obtained. Finally, Section 5 provides some conclusions and ongoing challenges.

## 2 Related Work

The research in Intrusion Detection began in the 1980s when Anderson suggested that the normal behavior of a user could be characterized analyzing his/her usual set of event logs. Since then, the area has attracted a significant number of researchers.

The first application to detect 'unusual' events or attacks was named IDES (Intrusion Detection Expert System) and it was developed by Dorothy Denning [2]. The basic idea of such a system was to monitor the normal activity in a mainframe and based on those activities define a set of rules which would allow the detection of anomalies.

It is worth mentioning that currently not only the core of the problem keeps being the same, but the complexity of the systems has increased considerably. Whereas Denning's approach suggested to analyze the event logs of a mainframe where the users were connected to, currently a system is composed by a lot of servers and workstations where each one creates its own event logs.

More systems that used data mining algorithms on event logs were proposed, but all them were based on centralized Unix events. In [3] a method for discovering temporal patterns in event sequences was proposed. Debar et al. proposed a system which could analyze the behavior of user activity using neural networks [4]. Neural networks were also used for anomaly detection based on Solaris BSM (Basic Security Module) audit data [5]. Lee and Stolfo used in [6] audit data from Unix machines to create behavior patterns using association rules and frequent episode mining, this way a set of events that occurred in a given time window could be discovered. In [7] Lane investigated the use of Hidden Markov Models for user pattern generation.

The source of the event logs used turns as the main difference with our proposed work. Almost the 90% of the population uses Windows systems, and the events are stored in each host. The complexity of centralizing and analyzing this information increases significantly. Also, our approach focuses on discovering the behavior of the hosts, and not the users related to them. This way we do not focus only on the usage patterns for intrusion detection, but more on any anomalous behavior that could happen (i.e. misconfigurations).

In order to allow the centralization of all this information and make easier the use of it, Security Information Management (SIM) tools have been developed. Currently, there are many applications developed with the purpose of detecting unusual behaviors. Tools such as Tripwire[1], Samhain [2], GFI EventsManager [3]

---

[1] Tripwire: http://sourceforge.net/projects/tripwire/
[2] Samhain: http://www.samhain.org/
[3] GFI Events Manager: http://www.gfi.com/es/eventsmanager/

and specially OSSEC [4] and Cisco MARS (Cisco Security Monitoring, Analysis, and Response System) [5] are an example of it. Nevertheless, only GFI EventsManager, OSSEC and Cisco MARS can be used in Windows environments and their strategies to analyze need to be improved. These tools, except Cisco MARS, are mainly focused on monitoring modifications in configuration, administration actions, identification of system errors and suspicious security problems. But, neither of them has the ability to generate sequential models which allow to detect unusual events. In this sense, different approaches have tried to discover the correlation between events [8]. Even some of them have worked with summarized data [9]. Specific tools for mining event logs have also been developed [10]. Other options that have been studied are the use of techniques used in temporal series mining [11] or the use of techniques for mining frequent itemsets [12].

It is clear the need of a system which clusters logically the security event logs generated in Windows systems. Therefore in the following sections we describe an approach to classify and correlate such events so that they can be used for further applications.

## 3 Analysis of Windows Security Event Logs

Windows classifies the events in different categories that are stored in independent records, such as System Registry, Application Registry, DNS Registry and Security Registry. This paper focuses on the events stored in the security registry, such as session logons or changes of privileges. It can be activated from the Administrator of domain users (NT) or security guidelines (W2K, W2K3) and it is available in all the versions of Windows Professional and Server.

Each event contains information like type of event, date and time information, event source (the software that has registered the event), category, event that has been produced (event ID), user who has produced and station where the event has occurred. Finally, Windows allows to define nine different categories related to security events.

- Account logon events: This event defines the authentication of a user from the point of view of the system. A single event of this type is not very meaningful but if there are many attempts in a short period of time, it can mean a scan activity or brute force attack.
- Account management: Activity related to the creation, management and delete of individual user accounts or groups of users.
- Directory service access: Access to any object that contains System Access Control Lists (SACL).
- Logon events: User authentication activity coming from local station as well as from the system that triggered the activity in a network.
- Object access: Access to file system and objects of the registry. It provides an easy to use tool to register changes in sensible files.

---

[4] OSSEC: http://www.ossec.net/
[5] Cisco MARS: http://www.cisco.com/en/US/products/ps6241/

- Policy changes: Changes in the access policy and some other modifications.
- Privilege use: Windows allows to define granular permissions to carry out specific tasks.
- Process tracking: It generates detailed information about when a process starts and finishes or when the programs are activated.
- System events: It registers information that affects the integrity of the system.

In this work, we are going to consider events generated by 4 different Domain Controllers (DC) during 4 days. From this point on, these servers will be named as Alfa, Beta, Gamma and Delta. Table 1 shows the number of events generated by each station each day. It is worth mentioning that the Gamma server generates much more events than the rest of the DCs. Moreover, the more events the system generates, more complex is their analysis. That is why the data mining techniques seem a promising approach for this type of data.

**Table 1.** Number of events to be analysed in the experiment

|  | Day 1 | Day 2 | Day 3 | Day 4 | Total |
|---|---|---|---|---|---|
| *Gamma* | 4.811.036 | 2.957.733 | 3.767.927 | 1.085.619 | 12.622.315 |
| *Beta* | 499.589 | 881.758 | 876.110 | 895.249 | 3.152.706 |
| *Delta* | 77 | 66 | 78 | 105 | 326 |
| *Alfa* | 1.565.283 | 1.492.202 | 1.540.150 | 1.996.107 | 6.593.742 |

## 4 Clustering Event Sources

In this section we are going to describe the experiments carried out using Windows event logs. For that, we have followed the usual steps suggested in any Data Mining process [13].

### 4.1 Learning the Application Domain

The event logs have some special features that have to be taken into account in the clustering process. For that, firstly, the dataset is analyzed, extracting descriptive statistics of each attribute. Statistics only show the number and the percentage of different values for each attribute. Usefulness of each attribute was defined by the distribution of its values. All those attributes where more than 80% of the events belonged to the same value were ruled out. Those attributes that were statistically dependant on any other actions were ruled out too (for instance Message vs EventID).

After analyzing the data we realized that although there were 22,369,089 events, the number of different type of events (different EventID-s) was 28. We decided to analyze the events generated by each server, ruling out all the attributes except Workstation name, Event ID, User ID and Timestamp.

## 4.2 Feature Selection

The attribute 'Event ID' is the key feature when it comes to carry out the analysis. It means that the statistics that are going to be used as input will be classified based on such a feature.

This step of the process is critical and may influence directly the results we obtain. Statistics are proposed as those indicators that might be key to express computer behavior based on security event logs. After analyzing the information the following features were identified in order to cluster sources of Windows logs.

1. Number of total events ($num\_id$)
2. Number of different types of events ($num\_diff\_id$)
3. Number of different users ($num\_diff\_user$)
4. Most frequent event ($freq\_event\_1$)
5. Second most frequent event ($freq\_event\_2$)
6. Percentage of events equal to the most frequent event ($perc\_event\_1$)
7. Percentage of event equal to the second most frequent event ($perc\_event\_2$)
8. Most frequent event in the most active second ($freq\_event\_max\_sec$)
9. Most frequent event in the most active minute ($freq\_event\_max\_min$)
10. Event of the most largest sequence of the same event ($long\_event\_id$)
11. Length of the most largest sequences of the same event ($long\_event\_num$)

## 4.3 Application of Clustering Techniques

Once the attributes have been selected, two different clustering processes have been carried out.

**Clustering of statistic data using K-means.** Clustering is a data mining technique which groups similar instances based on the similarities of their attributes. The basic idea is to minimize the distance between the instances of the same cluster and maximize the distance between different clusters. There are many different clustering techniques such as hierarchical clustering or partitional clustering. In our case, the simplest approach (K-means) seems to be enough. One particularity of K-means is that it is necessary to give the number of clusters to discover in advance. In this work, with the aim of obtaining patterns of the different machines, this constant is known, i.e. 4 in our case.

K-means technique [14] selects K points as initial centroids of the clusters. Then it assignees all instances to the closest centroid and it re-computes the centroid of each cluster. This process is repeated until the centroids of clusters remain in the same position.

We have applied such a technique to the data collected from different events and summarized in Table 1. We know in advance that the first four instances belong to events occurred during four days in the station named Alfa, the following four instances belong to Beta station and so on. The application of the K-means technique on the selected attributes ($num\_id$, $num\_diff\_id$ and $long\_event\_num$ in our case) provided as result four clusters, which match with the four servers analyzed.

**Discovering frequent event sequences.** So far, we have considered the events as independent events and we have analyzed them from a statistical point of view. The events we are considering in this work are the following ones:

- 538; User Logoff
- 540; Successful Network Logon
- 576; Special privileges assigned to new logon
- 578; Privileged object operation

If we order the events based on their timestamps, we will get a sequence of events, which can be analyzed in different ways. This second approach mainly focuses on the analysis of these 16 different sequences generated by the 4 DCs during 4 days. A sequence of events is a set of nominal symbols which indicates the occurrence of different events. Our work has focused on analysing what events usually interrupt previous events. Let us consider that the system has recorded the following sequence:

$$540 - 540 - 540 - 538$$

We could say that in this case, the event 540 (Successful Network Logon) has been interrupted by the event 538 (User Logoff). In that sense, we have considered all the possible interruptions, so that taking into account that we are considering 28 different events, we have generated a 28 x 28 matrix. In that matrix we store how many times an event has interrupted a previous event. Let us consider the example depicted in Figure 2. It means that 2500 times the event 540 has been interrupted by the event 538.

|  | | Interrupted by... | |
|---|---|---|---|
|  |  | 538 | Event y |
| Interrupted event | 540 | 2500 | ------ |
|  | Event x | ------ | ------ |

**Fig. 1.** Interruptions matrix

The content of such a matrix is represented by means of an array, where the first 28 values define the interruptions of the first event (in this case the event 538 User Logoff). Thus, the first value will mean how many times the 538 event is interrupted by itself (we will consider as 0), the second one how many times it is interrupted by the event 540 (Successful Network Logon), and so on.

After representing such values in an array, we depicted them in graphics where the graphic Alfa1 shows the interruptions for the Alfa server in the first day,

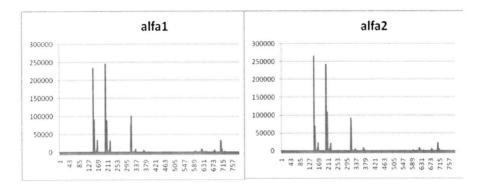

**Fig. 2.** Day 1 and 2 of Alfa server

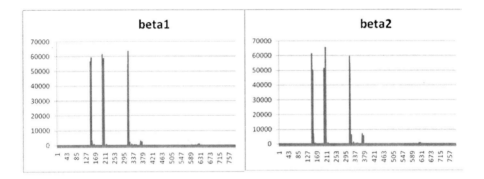

**Fig. 3.** Day 1 and 2 of Beta server

Alfa2 shows the interruptions of the same server in the second day and so on. The following pictures show the series obtained for the stations Alfa and Beta in the first two days.

Looking at the figures we realized that the results for a particular server in different days were very similar. Moreover, the dissimilarities with the rest of the servers could facilitate the clustering process. Thus, taking as starting point the 16 series (Alfa1, Alfa2, Alfa3, Alfa4, Beta1, Beta2, Beta3, Beta4, Gamma1, Gamma2, Gamma3, Gamma4, Delta1, Delta2, Delta3, Delta4) we carried out a clustering process using again the K-means technique.

In order to compare and therefore cluster the interruptions, we will need criteria to measure the similarity. Let us consider these two set of interruptions X and Y:

$$X = X_1, X_2, X_3, ... X_n \qquad (1)$$

$$Y = Y_1, Y_2, Y_3, ... Y_n \qquad (2)$$

Similarity between sets of interruptions will be given by the Manhattan distance between them D (X,Y):

$$D(X,Y) = \sum_{i=1}^{n} |X_i - Y_i| \qquad (3)$$

Table 2 shows the results of the clustering process. 15 out of 16 series were well classified, misclassifying only one series of the Gamma DC.

**Table 2.** Clustering of frequent event sequences

| Series number | Name of the series | Assigned Cluster |
|---|---|---|
| 1 | Alfa 1 | 2 |
| 2 | Alfa 2 | 2 |
| 3 | Alfa 3 | 2 |
| 4 | Alfa 4 | 2 |
| 5 | Beta 1 | 4 |
| 6 | Beta 2 | 4 |
| 7 | Beta 3 | 4 |
| 8 | Beta 4 | 4 |
| 9 | Gamma 1 | 1 |
| 10 | Gamma 2 | 1 |
| 11 | Gamma 3 | 1 |
| 12 | Gamma 4 | 2 |
| 13 | Delta 1 | 3 |
| 14 | Delta 2 | 3 |
| 15 | Delta 3 | 3 |
| 16 | Delta 4 | 3 |

## 5  Conclusions and Ongoing Challenges

Discovering frequent patterns in event logs is the first step to detect unusual behavior or anomalies. Besides proving that it is possible to detect patterns in event logs, different experiments have shown that different servers have different patterns and they can be found out and identified, even in Windows systems.

Thus, the experiments carried out at different stages have proved that the same server has very similar patterns during different days. In that sense, these experiments have been carried out with few Domain Controllers, so that it would be interesting to validate it with a larger set of servers and workstations.

Finally, it is worth to say that these results are work in progress that aims to detect anomalies in security event logs out of analyzing the event sources.

## References

1. Anderson, J.P.: Computer Security Threat Monitoring and Surveillance. Technical report, Fort Washington (1980)
2. Denning, D.E.: An Intrusion-Detection Model. IEEE transaction on Software Engineering 13(2), 222–232 (1987)

3. Teng, H., Chen, K., Lu, S.: Adaptive real-time anomaly detection using inductively generated sequential patterns. In: Proceedings of 1990 IEEE Computer Society Symposium on Research in Security and Privacy, Oakland, California, May 7-9, pp. 278–284 (1990)
4. Debar, H., Becker, M., Siboni, D.: A Neural Network Component for an Intrusion DetectionSystem. In: Proceedings, IEEE Symposium on Research in Computer Security and Privacy, pp. 240–250 (1992)
5. Endler, D.: Intrusion detection: Applying machine learning to solaris audit data. In: Proceedings of the 1998 Annual Computer Security Applications Conference (ACSAC 1998), Scottsdale, AZ, pp. 268–279. IEEE Computer Society, Los Alamitos (1998)
6. Lee, W., Stolfo, S.: Data Mining Approaches for Intrusion Detection. In: Proceedings of the Seventh USENIX Security Symposium (SECURITY 1998), San Antonio, TX (January 1998)
7. Lane, T., Brodley, C.E.: Temporal Sequence Learning and Data Reduction for Anomaly Detection. ACM Transactions on Information and System Security 2, 295–331 (1999)
8. Larosa, C., Xiong, L., Mandelberg, K.: Frequent pattern mining for kernel trace data. In: SAC 2008: Proceedings of the 2008 ACM symposium on Applied computing, Brazil, pp. 880–885 (2008)
9. Rana, A.Z., Bell, J.: Using event attribute name-value pairs for summarizing log data. In: AusCERT 2007 (2007)
10. Vaarandi, R.: Mining Event Logs with SLCT and LogHound. In: Proceedings of the 2008 IEEE/IFIP Network Operations and Management Symposium, pp. 1071–1074 (2008)
11. Viinikka, J.: Time series modeling for IDS Alert Management. In: ACM ASIAN Symposium on Information (2006)
12. Burdick, D., Calimlim, M., Gehrke, J.: A maximal frequent itemset algorithm for transactional databases. IEEE Trans. Knowl. Data Eng. 17(11), 1490–1504 (2005)
13. Fayyad, U., Piatetsky-Shapiro, G., Smyth, P.: The KDD process for extracting useful knowledge from volumes of data. Communications of the ACM 39(11), 27–34 (1996)
14. MacQueen, J.B.: Some Methods for classification and Analysis of Multivariate Observations. In: Proceedings of 5th Berkeley Symposium on Mathematical Statistics and Probability, vol. 1, pp. 281–297. University of California Press (1967)

# Text Clustering for Digital Forensics Analysis

Sergio Decherchi[1], Simone Tacconi[2], Judith Redi[1], Alessio Leoncini[1], Fabio Sangiacomo[1], and Rodolfo Zunino[1]

[1] Dept. Biophysical and Electronic Engineering, University of Genoa,
16145 Genova, Italy
{sergio.decherchi,rodolfo.zunino}@unige.it
[2] Servizio Polizia Postale e delle Comunicazioni
Ministero dell'Interno

**Abstract.** In the last decades digital forensics have become a prominent activity in modern investigations. Indeed, an important data source is often constituted by information contained in devices on which investigational activity is performed. Due to the complexity of this inquiring activity, the digital tools used for investigation constitute a central concern. In this paper a clustering-based text mining technique is introduced for investigational purposes. The proposed methodology is experimentally applied to the publicly available Enron dataset that well fits a plausible forensics analysis context.

**Keywords:** text clustering, forensics analysis, digital investigation.

## 1 Introduction

In the last years most of investigations performed by law enforcement agencies involve 'digital evidence', i.e. information and data of investigative value that is stored on, received, or transmitted by an digital device [1]. This evidence is acquired when data or electronic devices are seized. In this field, denominated 'digital forensics', due to increasing capability of mass storage devices, investigators have to face the problem of analysis of a great amount of information.

Key aspects of the investigational activities are the collection and analysis of available data. With this perspective digital data analysis [2-4] plays an important role in depicting a clearer vision of the context of interest.

The subsequent inquiry analysis is usually performed by a time-effort expensive human-based analysis: during this phase the analyst is requested to perform a heavy and complete study on the contents obtained from forensic acquisition.

During this activity, textual data (email, word processors etc…) constitute one of the core data sources that may contain relevant information. For this reason the typical requisite that emerges is a semi-automated texts contents analysis tool. As a consequence, in this research a two steps investigative process is proposed; these phases respectively are: text extraction and text clustering.

Text extraction is the process devoted to generate a collection of raw text file from information stored in digital devices. Text mining process relies on powerful tool to deal with large amounts of unstructured text data [5,6] possibly deriving from an investigational text extraction process.

The general area of text-mining methods comprises various approaches [6]: detection/tracking tools continuously monitor specific topics over time; document classifiers label individual files and build up models for possible subjects of interest; clustering tools process documents for detecting relevant relations among those subjects. As a result, text mining can profitably support intelligence and security activities in identifying, tracking, extracting, classifying and discovering patterns useful for building an investigative scenario.

This work addresses text clustering for forensics analysis based on a dynamic, adaptive clustering model to arrange unstructured documents into content-based homogeneous groups [7].

As benchmark the Enron emails database [8], provided the experimental domain. The research presented here shows that the document clustering framework [7] can find consistent structures suitable for investigative issues.

## 2 Textual Data Extraction

According to well known best practices of digital forensic, the first step of data extraction is the acquisition of data from devices, performed by means of a bit-stream copy, i.e. a copy of every bit of data, which includes the file slack and unallocated file space in which deleted files and e-mails are frequently recovered from.

Indeed, in the context of forensic analysis, it is common to involve deleted files, since these are often very interesting from the investigative point of view. Due to this reason, deleted file recovery constitutes the second phase of the process. For this purpose, there are two major strategies: a metadata-based and an application-based approach [9]. The first method is based on metadata of deleted files: technically speaking the related entry record of involved parent directories is recovered provided that such metadata still exist. If the metadata was reallocated to a new file or was wiped, an application based strategy is needed. In this case, chunks of data are searched for signatures that correspond to the header (start) and/or the footer (end) of known file types. This task is generally performed on the unallocated space of the file system; this stage allows also recovering files that are not pointed by any metadata, provided that their clusters were contiguously allocated. In this phase, obviously, one also extracts current files, i.e. files that are logically stored in the media.

The third phase, applied both to current files and to recovered deleted files is devoted to type identification and classification. This goal is not achievable by means of file extensions examination, since users can easily change them, but requires the analysis of headers and footers, applying to each file the same methodology of data carving.

The fourth phase is aimed to text extraction from files belonging to significant categories. In this stage, one may have both documental and non-documental files. In the case of documental files that are not purely textual, since text miner works on raw text files, a conversion is needed. For each non-documental file, if one wants to include this kind of files, one could extract their external metadata, residing the related entry record of parent director and/or their internal metadata, stored by software application inside the file itself. At this point, a collection of raw text files is ready to be further processed by the text mining tool.

## 3 Text Clustering

Text mining can effectively support analysis of information sources thanks to automatic means, which is of paramount importance to homeland security [10,11].

When applied to text mining, clustering algorithms are designed to discover groups in the set of documents such that the documents within a group are more similar to one another than to documents of other groups. The document clustering problem can be defined as follows. One should first define a set of documents $\mathcal{D} = \{D_1, \ldots, D_n\}$, a similarity measure (or distance metric), and a partitioning criterion, which is usually implemented by a cost function. In the case of flat clustering, one sets the desired number of clusters, Z, and the goal is to compute a membership function $\phi : \mathcal{D} \rightarrow \{1, \ldots, Z\}$ such that $\phi$ minimizes the partitioning cost with respect to the similarities among documents. Conversely, hierarchical clustering does not need to define the cardinality, Z, and applies a series of nested partitioning tasks which eventually yield a hierarchy of clusters.

### 3.1 Knowledge Base Representation

Every text mining framework should always be supported by an information extraction (IE) model [12,13] which is designed to pre-process digital text documents and to organize the information according to a given structure that can be directly interpreted by a machine learning system. Thus, a document D is eventually reduced to a sequence of terms and is represented as a vector, which lies in a space spanned by the dictionary (or vocabulary) $\mathcal{T} = \{t_j; j = 1, \ldots, n_T\}$. The dictionary collects all terms used to represent any document D, and can be assembled empirically by gathering the terms that occurs at least once in a document collection $\mathcal{D}$; by this representation one loses the original relative ordering of terms within each document. Different models [9,10] can be used to retrieve index terms and to generate the vector that represents a document D. However, the vector space model [14] is the most widely used method for document clustering. Given a collection of documents $\mathcal{D}$, the vector space model represents each document D as a vector of real-valued weight terms $\mathbf{v} = \{w_j; j=1,\ldots,n_T\}$. Each component of the $n_T$-dimensional vector is a non-negative term weight, $w_j$, that characterizes the $j$-th term and denotes the relevance of the term itself within the document D. In the following, $\mathcal{D} = \{D_u; u = 1,\ldots,n_D\}$ will denote the corpus, holding the collection of documents to be clustered. The set $\mathcal{T} = \{t_j; j = 1,\ldots, n_T\}$ will denote the vocabulary, which is the collection of terms that occur at least one time in $\mathcal{D}$ after the pre-processing steps of each document $D \in \mathcal{D}$ (e.g., stop-words removal, stemming [12]).

### 3.2 Clustering Framework

The clustering strategy is mainly based on two aspects: the notion of distance between documents and the involved clustering algorithm.

According to [7] the used distance consists in a weighted Euclidean distance plus a term based on stylistic information [7]. Defined as $\alpha$ the weight, $\Delta^{(f)}$ the Euclidean term and $\Delta^{(b)}$ the stylistic term, then the distance between $D_u$ and $D_v$ can be worked out as:

$$\Delta(D_u, D_v) = \alpha \cdot \Delta^{(f)}(D_u, D_v) + (1-\alpha) \cdot \Delta^{(b)}(D_u, D_v), \tag{1}$$

Strictly speaking (1) is not a metric space because does not guarantee the triangular inequality, for this reason (1) can be more properly considered a similarity measure of data. This distance measure has been employed in the well known Kernel K-Means [7] clustering algorithm.

The conventional k-means paradigm supports an unsupervised grouping process [15], which partitions the set of samples, $\mathcal{D} = \{D_u; u = 1,...,n_D\}$, into a set of Z clusters, $C_j$ ($j = 1,..., Z$). In practice, one defines a "membership vector," which indexes the partitioning of input patterns over the K clusters as: $m_u = j \Leftrightarrow D_u \in C_j$, otherwise $m_u = 0$; $u = 1,..., n_D$. It is also useful to define a "membership function" $\delta_{uj}(D_u, C_j)$, that defines the membership of the $u$-th document to the $j$-th cluster: $\delta_{uj} = 1$ if $m_u = j$, and 0 otherwise. Hence, the number of members of a cluster is expressed as

$$N_j = \sum_{u=1}^{n_D} \delta_{uj}; \quad j = 1,..., Z; \tag{2}$$

and the cluster centroid is given by:

$$\mathbf{w}_j = \frac{1}{N_j} \sum_{u=1}^{n_D} \mathbf{x}_u \delta_{uj}; \quad j = 1,..., Z; \tag{3}$$

where $\mathbf{x}_u$ is any vector-based representation of document $D_u$.

The kernel based version of the algorithm is based on the assumption that a function, $\Phi$, can map any element, $D$, into a corresponding position, $\Phi(D)$, in a possibly infinite dimensional Hilbert space. In the new mapped space clustering centers become:

$$\Psi_j = \frac{1}{N_j} \sum_{u=1}^{n_D} \Phi_u \delta_{uj}; \quad j = 1,..., Z. \tag{4}$$

According to [7] this data mapping allows different salient features able to ease the clustering procedure.

The ultimate result of the clustering process is the membership vector, **m**, which determines prototype positions (4) even though they cannot be stated explicitly. As per [7], for a document, $D_u$, the distance in the Hilbert space from the mapped image, $\Phi_u$, to the cluster $\Psi_j$ as per (4) can be worked out as:

$$d(\Phi_u, \Psi_j) = \left\| \Phi_u - \frac{1}{N_j} \sum_{v=1}^{n_D} \Phi_v \right\|^2 = 1 + \frac{1}{(N_j)^2} \sum_{m,v=1}^{n_D} \delta_{mj} \delta_{vj} \Phi_m \cdot \Phi_v - \frac{2}{N_j} \sum_{v=1}^{n_D} \delta_{vj} \Phi_u \cdot \Phi_v. \tag{5}$$

By using expression (5), one can identify the closest prototype to the image of each input pattern, and assign sample memberships accordingly.

## 4 Forensic Analysis on Enron Dataset

In this study, for simulating an investigational context, Enron emails dataset [8] was used. The consequence of this choice is that textual data extraction process is not explicitly performed in these experiments: this aspect does not compromise the correctness of the overall proposed approach; extracting real data from on-field devices is not an easy task due to privacy issues; for these reasons the publicly available Enron emails dataset was used.

The Enron email dataset [8] provides a reference corpus to test text-mining techniques that address investigational applications [2-4]. The Enron mail corpus was posted originally on Internet by the Federal Energy Regulatory Commission (FERC) during its investigation on the Enron case. FERC collected a total of 619,449 emails from 158 Enron employees, mainly senior managers. Each message included: the email addresses of the sender and receiver, date, time, subject, body and text. The original set suffered from document integrity problems, hence an updated version was later set up by SRI International for the CALO project [16]. Eventually, William Cohen from Carnegie Mellon University put the cleaned dataset online [8] for researchers in March 2004. Other processed versions of the Enron corpus have been made available on the web, but were not considered in the present work because the CMU version made it possible fair comparison of the obtained results with respect to established, reference corpora in the literature.

Five employees were randomly selected: *White S., Smith M., Solberg G., Ybarbo P. and Steffes J.* Collected emails (see tab.1) were separately processed, thus obtaining five different scenarios for each employee.. The underlying hypothesis was that email contents might also be characterized by the role the mailbox owner played within the company. Toward that end, when applying the clustering algorithm, only the 'body' sections of the emails were used, and sender/receiver, date/time info were discarded.

Table 1. Names and corresponding number of Emails

| Name | Number of Emails |
|---|---|
| White S. | 3272 |
| Smith M. | 1642 |
| Solberg G. | 1081 |
| Ybarbo P | 1291 |
| Steffes J | 3331 |

The performed experiments used a number of 10 clusters: this choice was guided by the practical demand of obtaining a limited number of informative groups. Tables from 2 to 7 report on the results obtained by these experiments: it shows the terms that characterize each of the clusters provided by the clustering framework for each employee. For each cluster, the most descriptive words between the twenty most frequent words of the cluster are listed; reported terms actually included peculiar abbreviations: "ect" stands for Enron Capital & Trade Resources, "hpl" stands for Houston Pipeline Company, "epmi" stands for Enron Power Marketing Inc, "mmbtu"

**Table 2.** *Smith* results

| Cluster ID | Most Frequent and Relevant Words |
|---|---|
| 1 | employe, business, hotel, houston, company |
| 2 | pipeline, social, database, report, link, data |
| 3 | ect, enronxg |
| 4 | coal, oil, gas, nuke, west, test, happi, business |
| 5 | yahoo, compubank, ngcorp, dynegi, night, plan |
| 6 | shank, trade |
| 7 | travel, hotel, continent, airport, flight, sheraton |
| 8 | questar, paso, price, gas |
| 9 | schedule, london, server, sun, contact, report |
| 10 | trip, weekend, plan, ski |

**Table 3.** *Solberg* results

| Cluster ID | Most Frequent and Relevant Words |
|---|---|
| 1 | paso, iso, empow, ub, meet |
| 2 | schedule, detected, california, iso, parsing |
| 3 | ub, employe, epe, benefit, contact, ubsq |
| 4 | shcedule, epmi, ncpa, sell, buy, peak, energi |
| 5 | dbcaps97, data, failure, database |
| 6 | trade, pwr, impact, london |
| 7 | awarded, california, iso, westdesk, portland |
| 8 | error, pasting, admin, sql, attempted |
| 9 | failure, failed, required, intervention, crawl |
| 10 | employe, price, ub, trade, energi |

stands for Million British Thermal Units, "dynegi" stands for Dynegy Inc, a large owner and operator of power plants and a player in the natural gas liquids and coal business, which in 2001 made an unsuccessful takeover bid for Enron.

From the investigational point of view some interesting aspects emerges:

In *Smith* results there is an interesting cluster (cluster 10) in which the context seems not strictly adherent to the workplace usual terms. This means that analyzing that bunch of email may mean acquiring sensible private life information potentially useful for investigation.

*Solberg* results do not underline any particular peculiarity. However it is curious to observe that his emails are full of server errors.

**Table 4.** *Steffes* results

| Cluster ID | Most Frequent and Relevant Words |
|---|---|
| 1 | ferc, rto, epsa, nerc |
| 2 | market, ferc, edison, contract, credit, order, rto |
| 3 | ferc, report, approve, task, imag, attach |
| 4 | market, ee, meet, november, october |
| 5 | california, protect, attach, testimoni, whshington |
| 6 | stock, billion, financial, market, trade, investor |
| 7 | market, credit, ee, energi, util |
| 8 | attach, gov, energy, sce |
| 9 | affair, meet, report, market |
| 10 | gov, meet, november, imbal, pge, usbr |

**Table 5.** *White* results

| Cluster ID | Most Frequent and Relevant Words |
|---|---|
| 1 | meet, chairperson, oslo, invit, standard, smoke |
| 2 | confidential, attach, power, internet, copy |
| 3 | West, ect, meet, gas |
| 4 | gopusa, power, report, risk, inform, managment |
| 5 | webster, listserv, subscribe, htm, blank, merriam |
| 6 | report, erv, asp, efct, power, hide |
| 7 | ect, rhonda, john, david, joe, smith, michae,l mike |
| 8 | power |
| 9 | mvc, jpg, attach, meet, power, energy, canada |
| 10 | calendard, standard, monica, vacation, migration |

**Table 6.** *YBarbo* results

| Cluster ID | Most Frequent and Relevant Words |
|---|---|
| 1 | report, status, week, mmbtu, price, lng, lpg, capacity |
| 2 | tomdd, attach, ship, ect, master, document |
| 3 | london, power, report, impact, gas, rate, market, contact |
| 4 | dpc, transwestern, pipeline, plan |
| 5 | inmarsat, galleon, eta, telex, master, bar, fax, sea, wind |
| 6 | rate, lng, price, agreement, contract, meet |
| 7 | report, houston, dubai, dial, domest, lng, passcode |
| 8 | power, dabhol, india, dpc, mseb, govern, maharashtra |
| 9 | cargo, winter, gallon, price, eco, gas |
| 10 | arctic, cargo, methan |

*Steffes* results seem extremely interesting. Cluster 6 underlines a compact cluster in which important financial aspects of Enron group are discussed. In particular some key words as F.E.R.C. (that stands for Federal Energy Regulatory Commission) are particularly expressive.

From *White* frequent terms analysis one can observe that cluster 2 contains several time the word "*confidential*" making this group interesting and worth of further analysis. Cluster 7 exhibits a strange content; it is characterized completely by names of people. This could indicate that these emails may concern private life.

*YBarbo* emails have no particular feature. The only aspect that can be understood is that his position is tightly linked to international affairs.

## References

1. U.S. Department of Justice, Electronic Crime Scene Investigation: A Guide for First Responders, I Edition, NCJ 219941 (2008), http://www.ncjrs.gov/pdffiles1/nij/219941.pdf
2. Chen, H., Chung, W., Xu, J.J., Wang, G., Qin, Y., Chau, M.: Crime data mining: a general framework and some examples. IEEE Trans. Computer 37, 50–56 (2004)
3. Seifert, J.W.: Data Mining and Homeland Security: An Overview. CRS Report RL31798 (2007), http://www.epic.org/privacy/fusion/crs-dataminingrpt.pdf
4. Mena, J.: Investigative Data Mining for Security and Criminal Detection. Butterworth-Heinemann (2003)
5. Sullivan, D.: Document warehousing and text mining. John Wiley and Sons, Chichester (2001)
6. Fan, W., Wallace, L., Rich, S., Zhang, Z.: Tapping the power of text mining. Comm. of the ACM 49, 76–82 (2006)
7. Decherchi, S., Gastaldo, P., Redi, J., Zunino, R.: Hypermetric k-means clustering for content-based document management. In: First Workshop on Computational Intelligence in Security for Information Systems, Genova (2008)
8. The Enron Email Dataset, http://www-2.cs.cmu.edu/~enron/
9. Carrier, B.: File System Forensic Analysis. Addison-Wesley, Reading (2005)
10. Popp, R., Armour, T., Senator, T., Numrych, K.: Countering terrorism through information technology. Comm. of the ACM 47, 36–43 (2004)
11. Zanasi, A. (ed.): Text Mining and its Applications to Intelligence, CRM and KM, 2nd edn. WIT Press (2007)
12. Manning, C.D., Raghavan, P., Schütze, H.: Introduction to Information Retrieval. Cambridge University Press, Cambridge (2008)
13. Baeza-Yates, R., Ribiero-Neto, B.: Modern Information Retrieval. ACM Press, New York (1999)
14. Salton, G., Wong, A., Yang, L.S.: A vector space model for information retrieval. Journal Amer. Soc. Inform. Sci. 18, 613–620 (1975)
15. Linde, Y., Buzo, A., Gray, R.M.: An algorithm for vector quantizer design. IEEE Trans. Commun. COM 28, 84–95 (1980)
16. Bekkerman, R., McCallum, A., Huang, G.: Automatic Categorization of Email into Folders: Benchmark Experiments on Enron and SRI Corpora. CIIR Technical Report IR-418 (2004), http://www.cs.umass.edu/~ronb/papers/email.pdf

# A Preliminary Study on SVM Based Analysis of Underwater Magnetic Signals for Port Protection

Davide Leoncini[1], Sergio Decherchi[1], Osvaldo Faggioni[2], Paolo Gastaldo[1], Maurizio Soldani[2], and Rodolfo Zunino[1]

[1] Dept. Biophysical and Electronic Engineering, University of Genoa,
16145 Genoa, Italy
{davide.leoncini,sergio.decherchi,paolo.gastaldo}@unige.it
rodolfo.zunino@unige.it
[2] Istituto Nazionale di Geofisica e Vulcanologia,
19025 Portovenere (SP), Italy
{osvaldo.faggioni,maurizio.soldani}@ingv.it

**Abstract.** People who attend to the problem of underwater port protection usually use sonar based systems. Recently it has been shown that integrating a sonar system with an auxiliary array of magnetic sensors can improve the effectiveness of the intruder detection system. One of the major issues that arise from the integrated magnetic and acoustic system is the interpretation of the magnetic signals coming from the sensors. In this paper a machine learning approach is explored for the detection of divers or, in general, of underwater magnetic sources that should ultimately support an automatic detection system which currently requires a human online monitoring or an offline signal processing. The research proposed here, by means of a windowing of the signals, uses Support Vector Machines for classification, as tool for the detection problem. Empirical results show the effectiveness of the method.

**Keywords:** underwater detection systems, port protection, magnetic signal processing, Support Vector Machine.

## 1 Introduction

For many years security has not been perceived by people as a necessity. Today, after some dramatic events such as September 11 2001, security issue has become a serious concern not only for governments. In this scenario the importance of physical security has increased; in particular, during the last five years, the research concerning underwater port protection has made some substantial achievements [1, 3-6].

First of all the target of underwater intruder detection systems has been extended from a military one, such as an enemy nation navy submarine, to a terrorist one, such as a diver intruder. This produced a secondary effect concerning the up to date of the technology used to detect underwater sources: traditional sonar systems resulted to be insufficient to solve this task, bringing back importance to magnetic based systems [4, 5]. The analysis and comparison of the performances of the two different approaches point out their peculiarities: acoustic arrays guarantee optimum volumetric control but lack in peripheral surveillance; vice versa magnetic subsystems achieve high peripheral security performances but partially fail in volumetric control. These considerations suggest the integration of both detection approaches into a dual system [6].

This integration guarantees a good effectiveness to the complete system: overlapping of the acoustic and magnetic subsystems supplies shadow areas avoidance and consequently prevents possible intrusions. Moreover in the zone of maximum uncertainty of each method the lack in performance of one approach is counterbalanced by the co-occurring presence of the other cooperating subsystem. While acoustic systems today are a commercial reality, magnetic underwater surveillance is still an open research field.

These premises lead to the demand of proper tools able to analyze the magnetic subsystem output. Beside classical analysis techniques [1, 3] the purpose of this paper is introducing a machine learning tool, Support Vector Machine for classification, as a possible approach to automate the detection of diver intrusion patterns on the supplied data. In particular, machine learning techniques have been already successfully used when coping with sonar signals [7]; here the purpose is showing that an analogous approach can be also carried when dealing with magnetic signals. Section 2 introduces the magnetic subsystem architecture while Section 3 exposes SVM theory, data extraction and experimental results.

## 2 The "MACmag" Magnetic Subsystem

Nowadays magnetic sensors have extremely high sensitivities and are able, in theory, to detect signals generated by divers without any problem. This capability is strongly compromised in practice by the spectral content of the Earth's magnetic field in high noise environments, such as port areas, characterized by an extremely wide band and high amplitude components, which often hide the target signal. Given $M$ spectral components of the magnetic field, if we call $E_i$ the energy associated with the i-th component, the information content $Q$ is given by [2]:

$$Q = \sum_{i=1}^{M} E_i \tag{1}$$

Whereas the information capacity $C_i$, that is the capacity associate to the i-th elementary spectral component with its physical generator, is given by the ratio between the energy $E_i$ and the total energy in which it is contained:

$$C_i = \frac{E_i}{Q} \tag{2}$$

The range of value of $C_i$ is between 1 (monochromatic signal) and 0 (white noise or insufficient target signal amplitude):

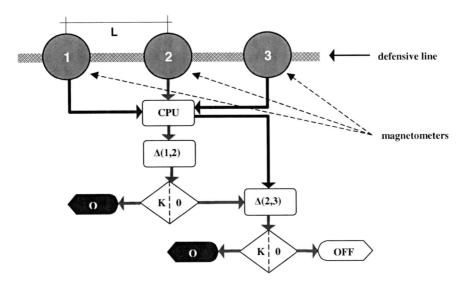

**Fig. 1.** Operative structure of the elementary cell of the MAC mag subsystem

$$\lim_{Q \to E_i} C_i = 1 \qquad \lim_{Q \to \infty} C_i = 0 \qquad \lim_{E_i \to 0} C_i = 0 \qquad (3)$$

Given two magnetometers, one as sentinel and the other as reference, to protect a critical area, one indicates with $N$ the noise measured by both the magnetometers. By $T$ is indicated the target signal acquired only by the sentinel magnetometer. As shown in [3] it can be stated that the sentinel listen to $N+T$ and the reference measures the environmental noise $N$. This configuration can be obtained using two different architectures of the magnetic subsystem: the first employs the magnetic field acquired from the previous or next sensor in the array as noise reference (so that each instrument in the array operates both as sentinel and as reference) and is known as SIMAN-type network (Self-referred Integrated MAgnetic Network); the second is based on a sensor array and another external device used to obtain noise reference values (so that all the instruments in the array operate only as sentinel) and is called RIMAN-type network (Referred Integrated MAgnetic Network) [4, 6]. The system employed in the present work consists of two magnetometers in a SIMAN configuration. However, this configuration does not represent a full operational unit of the SIMAN network; a diver crossing halfway between the two sentinel magnetometers induces an analogous signal in both the devices and, consequently, this produces the target signal removal if target and reference signal are subtracted. Therefore, a full operational unit needs a third magnetometer which allows a comparison $\Delta(1,2)$, between the first pair of sensors, and $\Delta(1,3)$, between the second pair, such that the removal of the target can occur for at most one pair only (see Fig. 1) but not for the whole system.

Nevertheless the experimental configuration employed, including the two magnetometers, is clearly suitable for experimental validation of the MACmag component, with the exclusion of target crossings halfway between the two sensors. The magnetic signal used in ours experiments has been grabbed in this way from the sentinel and reference sensors in noisy environmental conditions and considering a civil diver as target. The eventual effectiveness of this architecture in detecting divers lies in this reference-target technique as exposed in [5]. The exigency of an automated detection system leads to the following explorative machine learning based analysis.

## 3 Support Vector Machines for Classification

Support Vector Machines (SVM) constitutes a robust and well known classification algorithm [8]. The good classification performance of SVMs is due to the concept of margin maximization, whose roots are deeply connected with Statistical Learning Theory [8]. As usual in learning machines, SVM has a learning phase and a prediction phase. In the learning stage the machine sees the training patterns and learns a rule (an hyperplane) able to separate data in two groups according to data labeling. Conversely in the forward (prediction) phase the machine is asked to predict labels of new and unseen patterns.

From the formal point of view the following notation will be used:

- $n_p$ is the number of patterns used as training set
- $\mathbf{X}$ is the training set
- $\mathbf{x} \in R^{n_i}$ is a pattern belonging to $\mathbf{X}$ where $n_i$ is the data dimensionality
- $f(\mathbf{x}) = sign(\mathbf{wx} + b)$ is the prediction function based on the hyperplane defined by the normal $\mathbf{w}$ and the bias $b$
- $\mathbf{y}$ is the vector of labels of the training set, with $\mathbf{y} \in \{-1, 1\}$

Given these definitions the cost function to be minimized for obtaining optimal $\mathbf{w}$ and $b$ is:

$$C \sum_{i=1}^{n_p} (1 - y_i(\mathbf{wx} + b))_+ + \frac{1}{2}\|\mathbf{w}\|^2 \qquad (4)$$

Where the positive constant $C$ controls the tradeoff between data fitting (the first term) and regularization (the second term that represents margin maximization), and where $(k)_+$ indicates $\max(0, k)$.

Problem (4) can be solved via quadratic optimization algorithms; despite this fact, problem (4) is usually solved using its Lagrange dual formulation. The dual formulation makes possible to use non-linear mapping functions called kernel functions [8] that lead to non linear separating surfaces (see Fig. 2). This operation is possible observing that the only operations in which data are directly involved are dot products.

Calling $K(\mathbf{x}_l, \mathbf{x}_m)$ the kernel dot product between $\mathbf{x}_l$ and $\mathbf{x}_m$ it can be shown [8] that the dual problem of (4) is:

$$\min_\alpha \left\{ \frac{1}{2} \sum_{l,m=1}^{n_p} \alpha_l \alpha_m y_l y_m K(\mathbf{x}_l, \mathbf{x}_m) - \sum_{l=1}^{n_p} \alpha_l \right\} \text{ subject to: } \begin{cases} 0 \leq \alpha_l \leq C, \forall l \\ \sum_{l=1}^{n_p} y_l \alpha_l = 0 \end{cases} \qquad (5)$$

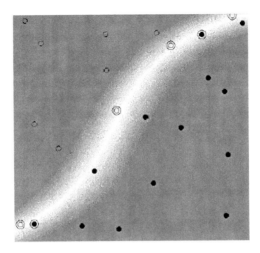

**Fig. 2.** Non linear separating surface obtained by SVM using a non linear kernel function

Where vector **α** is of length $n_p$ and represents the set of dual variables. Problem (5) poses the major problem of its optimization. To this regard fast optimization techniques has been developed [9]. One of these techniques, called Sequential Minimal Optimization [10], is the one that will be used for the following experimental section.

Once (5) has been optimized, as a final step, one has an efficient way to compute the bias $b$ [10]. Finally, provided **α** and $b$ the non linear prediction function can be written as:

$$f(\mathbf{x}) = sign\left(\sum_{i=1}^{n_p} \alpha_i y_i K(\mathbf{x}, \mathbf{x}_i) + b\right) \quad (6)$$

## 4 Experimental Results

The first addressed step is the definition of a suitable dataset for SVM based classification. Elaborated data refer to the problem of detecting the presence of a diver (class +1) or its absence (class -1).

Two quantities must be defined: the vector data **x** and its corresponding label $y$. The vector **x** can be created by windowing the signals coming from the magnetic subsystem: in particular given the original signal of length $m$, for each sample a window of width $l$ is grabbed. This means that the total number of windows (superposition of windows is allowed) is $m$-$l$. Because the signals coming from the subsystem are two (reference ambient signal and target detection signal), for each produced window the final pattern is built up by the concatenation of the two windows derived from the two signals.

The meaning of this concatenation of signals is showing to the machine contemporaneously the detection signal and the underlying environmental noise. In this way noise is not pre-filtered by a filtering system whose cut frequency choice is in general critical. For these reasons, and for the sake of simplicity, it is sufficient to work on time and not in frequency.

This translates in having $m$-$l$ patterns **x** each of size $2l$. Using $l=100$ the number of produced patterns is considerable; for this reason a sub-sampling technique has been employed. To obtain a meaningful dataset the sections of the signal which are characterized by an intrusion have been more densely windowed than the sections in which no intrusion occurs (see Fig 3).

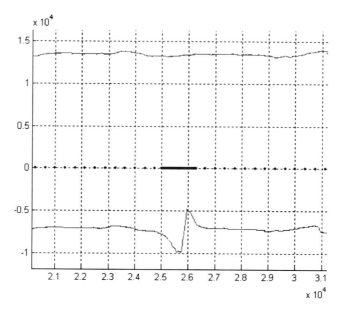

**Fig. 3.** The upper signal is the reference signal; the lower signal is the target signal. Dotted line in the middle represents the windowing density.

Table 1 summarizes the statistics of training and test data after the above mentioned sub-sampling technique.

**Table 1.** Dataset overview

| DataSet | Class +1 | Class -1 |
| --- | --- | --- |
| Training Set | 142 | 145 |
| Test Set | 144 | 150 |

After this preliminary step all data were normalized for each attribute in the domain [-1, +1]. The experimental session was carried by using a SVM with standard linear kernel [8] and SMO [10] optimizer. In particular the accuracy on the optimality conditions was set to 1e-3, a typical value for SVM training convergence (Karesh Kuhn Tucker conditions [8, 10]). The model was selected according to the $C$ regularization constant (as per (4)) that led to the lowest test set error.

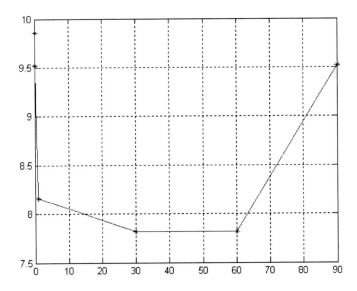

**Fig. 4.** Model Selection curve: x axis are C values, y axis are percentage error values

Figure 4 depicts the obtained curve for the C values {0.01, 0.1, 1, 30, 60, 90}; its shape is in accordance with theory [8], showing underfitting regions (small C values) and overfitting regions (big C values). The best performances are obtained with C = 30 and C = 60; for both an error of 7.82% occurs. Recalling that an underfitting behavior is usually preferable to an overfitting one [8], the final selected parameter C was set to C = 30. Confusion matrix in table 2 shows that despite the false negative rate is quite high, the false positive rate is quite reduced.

**Table 2.** Confusion matrix

| Predicted Value | Actual Value | |
|---|---|---|
| | Pos | Neg |
| Pos | 98.70% | 1.30% |
| Neg | 14.60% | 85.40% |

Considering that the current online detection system is human based or offline, the obtained results are quite promising mostly with the perspective of further improvements achievable by employing non linear kernel functions and data pre-processing techniques.

## 5 Conclusions

The current preliminary research showed that a Support Vector Machine for classification can be a feasible model for classification of underwater magnetic signals. A first experimentation with a linear kernel gave encouraging results about the achievable accuracy levels reachable with this approach; to get an on-field implementation, global accuracy and false negative rate must be further improved. Future works will deal with the optimal windows size, a deep SVM-based model selection and other supervised/unsupervised classifiers. Moreover other pre-processing techniques can be considered to ease the classification task.

# References

1. Faggioni, O., Gabellone, A., Hollett, R., Kessel, R.T., Soldani, M.: Anti-intruder port protection "MAC (Magnetic ACoustic) System": advances in the magnetic component performance. In: Proceedings of 1st WSS Conference, Copenhagen, Denmark, August 25-28 (2008) CD-ROM
2. Kanasewich, E.R.: Time Sequence Analysis in Geophysics. The University of Alberta Press, Edmonton (1981)
3. Faggioni, O., Soldani, M., Gabellone, A., Maggiani, P., Leoncini, D.: Development of anti intruders underwater systems: time domain evaluation of the self-informed magnetic networks performance. In: Corchado, E., Zunino, R., Gastaldo, P., Herrero, Á. (eds.) Proceedings of the International Workshop on Computational Intelligence in Security for Information Systems CISIS 2008. Advances in Soft Computin, vol. 53, pp. 100–107. Springer, Heidelberg (2009)
4. Gabellone, A., Faggioni, O., Soldani, M., Guerrini, P.: CAIMAN Experiment. In: Proceedings of UDT Europe 2007 Conference, Naples, Italy, June 5-7 (2007) CD-ROM
5. Gabellone, A., Faggioni, O., Soldani, M., Guerrini, P.: CAIMAN (Coastal Anti Intruder MAgnetometers Network). In: Proceedings of RTO-MP-SET-130 Symposium on NATO Military Sensing, Orlando, Florida, USA, March 12-14, NATO classified (2008) CD-ROM
6. Faggioni, O., Soldani, M., Zunino, R., Leoncini, D., Di Gennaro, E., Gabellone, A., Maggiani, P.V., Falcucci, V., Michelizza, E.: Building the Synthetic "MAC System": an Analytical Integration of Magnetic and Acoustic Subsystems for Port Protection Scenarios. In: Proceedings of UDT Europe 2009 Conference, Cannes, France, June 9-11 (2009) CD-ROM
7. Hettich, S., Bay, S.D.: The UCI KDD Archive. University of California, Department of Information and Computer Science, Irvine (1999), http://kdd.ics.uci.edu
8. Vapnik, V.: Statistical Learning Theory. Wiley-Interscience Pub., Hoboken (1998)
9. Schölkopf, B., Smola, A.J.: Learning with Kernels. MIT Press, Cambridge (2002)
10. Chang, C.C., Lin, C.J.: LibSVM: a library for Support Vector Machines, http://www.csie.ntu.edu.tw/~cjlin/papers/libsvm.pdf

# Fuzzy Rule Based Intelligent Security and Fire Detector System

Joydeb Roy Choudhury[1], Tribeni Prasad Banerjee[1], Swagatam Das[2],
Ajith Abraham[3], and Václav Snášel[4]

[1] Central Mechanical Engineering Research Institute, Embedded system Laboratory,
Durgapur-713209, India
[2] Electronics and Telecommunication Engineering Department Jadavpur University,
Jadavpur-700035, India
[3] Machine intelligence Research Labs (MIR Labs), USA
ajith.abraham@ieee.org
[4] VSB-Technical University of Ostrava, Czech Republic
vaclav.snasel@vsb.cz

**Abstract.** In fire alarm and monitoring system, fire detector has an interesting role to play with. But traditional fire detectors are unable to detect fire at its early state. Naturally, the braking out of fire cannot be controlled using such fire detectors. In this paper, we analyze the mechanism of fire-catching process, and implement by using a microprocessor based hardware and intelligent fire recognition software. In this paper, we also implement a fuzzy rule based intelligent early fire detection warning system. The early warning prior to the fault without any ambiguity can avoid the disaster against the fault taking some preventive measures.

**Keywords:** fire safety; intelligent recognition; early-fire detector; multi-sensor.

## 1 Introduction

Fire is a terribly destructive disaster among all kinds of disaster. In largest ten fire cases, most fires took place in public such as hospital, hotel, restaurant and wholesale supermarket. There is not only fault of supervision in fire fighting, but also mistake of no alarming in the early time of fire. And that caused that fire was not eliminated in the optimal opportunity and tragedies were resulted from those reasons.

At present, installation fire protection and fire detection and alarming system have been built inside most building. Though that system has important action in early fire detection and prevention fire, there have been serious problems of product quality, equipment supervision and alarm disposal in practice. And so all those reasons caused that normal function of equipment of fire detection and alarm and monitoring system is not exerted fully and influence severely the reliability of alarm [1,2]. Though intelligent detection technology and methods are applied in fire detector system and fire can be detected accurately, it is not suit to detect early fire because of self-shortcoming.

In this paper, a fuzzy rule based intelligent multi-sensor fire detector system has been introduces, one important aspect of a fuzzy logic is that it can operate on imprecise data having reasonably high information content. Thus using this logic the basic monitoring system can be converted to more qualitative than mere quantitative data based observation. A methodology has been described with a fuzzy logic based system using embedded processor for early warning system for a slow varying input. The proposed fire detector adopt multi-sensor and intelligent recognition program. And the proposed fire detector is of communication bus to data traffic. So this fire detector is of high precision and reliability.

## 2 Mechanism of Fire Occurred

Fire is of the burning phenomenon. In the process of fire, there is a chemical change, accompanied by the production of heat, light, smoke and flame. Expressional modality of ordinary combustible substance burning is that smoke and combustion gas is given off. Combustible substance burns in the condition of fully oxygen. Flame is produced accompanied by the production of visible and invisible light. And then temperature of circumstances or conditions is ascended because of releasing a great deal of heat. Process of fire usually is divided into four periods initial, smoldering developed and decay period [3, 4].

The proposed fire detector is just designed according to the principle mention above. And then in order to detector ensure that heat is detected to application.

## 3 Design of Intelligent Multi-sensor Fire Detector

In the initial stage, smog particles made beam or ray to scatter or reduce. Emitter ray or beam and receiver of ray are set in detection cell of smoke exit. Receiver element is connected to device of photoelectric converter. Electric signal is changed with change of ray intensity. Electric signal is amplified and input microprocessor. Fire alarm signal is sent out through intelligent recognition program. In the fire happening, Smog particles content of atmosphere will be increased rapidly. Smog particles are made mainly by smoke particles and CO gas. And temperature of fire happening will rise [5].

### 3.1 Work Principle of Temperature Sensor

Temperature is a fundamental physical parameter. When a fire is happening, the temperature may increase, and temperature sensor will vary with the circumstance and climate. Temperature change can be tested easily. DS18B20 is selected as temperature sensor. It belongs to one bus digital temperature sensor. Temperature signal is directly switched to serial digit. The volume of DS18B20 is small and low consumption.

DS18B20 is structured easily multi-point touring detection. The number of DS18B20 connected in one bus is not limited from theory. Because of DS18B20's unique product number, temperature of each point is measured and distinguished easily. For those reasons, position of fire may be ascertained when fire is happening. The survey range of DS18B20 is from -55 to +125. And 9-bit binary system represents temperature of each point. And using 16-bit fill code to read the temperature of each point by microprocessor. Fig.1 is the connection graph of mulit-DS18B20.

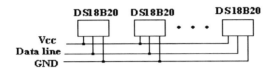

**Fig. 1.** Connection graph of Digital Temperature Sensor

## 3.2 Hardware Design of the Fire Detector

This system integrated with the previous system because only the alarm is give us the simple two values logic output (that is 0 (No-Fire),or 1(Fire ))this logic system for fault detection is not sufficient to detect a particular signal characteristics and its behavior at an early stage of the fault.[8,10] Once this two valued hard logic based approach is softened through grey scale based system - the behavior of the signal can be observed in a different way [11].

The hardware used for this developmental work is manufactured by Rigel Corporation USA and consists of one .R515JC board, which contains SIEMENS C515C processor with 64k byte on-chip ROM, 256 byte on-chip RAM and 2K byte of on-chip XRAM. Besides this board contains 32K of SRAM, 32K of monitor EPROM, and other necessary peripherals like battery back-up RTC (Real time clock), CAN(Controller area network) port, 3 full-duplex serial ports, etc.

The Rigel's FLASH package has been used to develop fuzzy algorithm. The Fuzzy Logic Application Software Helper (FLASH) is a commercial code generator developed by Rigel Corporation. FLASH editor is used for development of source code (Control Task Description File), which contains fuzzy rules. Finally FLASH generates assembly code from source code (Control Task Description File) to accomplish the control task. This assembly code generated by FLASH and the application software are merged and assembled using a simple absolute assembler (Please refer www.rigelcorp.com).

In addition to this a +5V DC is applied across the series combination of the thermistor (Negative temperature Coefficient) and a resistance $R_1$ as shown in the Figure1. When the heat source is kept near the thermistor, the resistance will decrease resulting in an increase of current. So, the voltage drop (Vi) across the resistance $R_1$ will increase. This voltage (Vi) is fed to input pin of the comparator. The reference input is varied in predefined manner through a pulse width modulated signal (PWM).

The reference input of the comparator (LM339) is connected to the out put port of the processor through a low pass filter for variable DC reference input generation and the out put of the comparator is connected to the external interrupt of the processor. The status indicator of the input condition of the signal is displayed to 3 LEDS with different grades as SAFE, UNSAFE AND DANGER (FIRE).

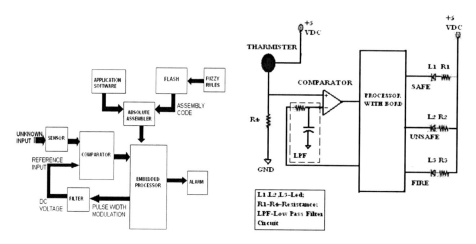

**Fig. 2.** Block Diagram of Hardware Module     **Fig. 3.** Hardware Module of Alarm

### 3.3 Software Design of the Fire Detector

In practice, an excellent fire detector depends on perfection hardware and function, besides further more recognition ability in fire circumstances. Fire signal and alarm are determined through temperature and smog particle change to implement. Fuzzy adjudication method is applied to the fire detector software design. Density change of smog particle and temperature change are set as adjudication input variable. "If, then" of fuzzy control[12], methods is merged to program thoughts. If temperature rise and density augment, then fire happening is adjudicated. If temperature rise and density is not changed, and temperature change ratio is swift, then fire is happening. If temperature and density are not changed, then fire is not happening. And so on. Not only is this fire-recognized method simple and time of recognized is short, but also is the capability of program occupied small [11].

In addition to, software design of the fire monitor alarm system is presented as module structure design program, the State diagram shown in figure 7. in order to debug conveniently and expand easily. Software program of fire detector mainly includes the following sub-systems: initialization program, self-examination program, fire detect itineration examination and disposal program, communication program, malfunction disposal program, keyboard and display sub-system program and so on. Otherwise in the course of software design, timer interrupt functions of microprocessor chip make the best use in the fire detector. Software clock is employed in the program design. In order to avoid entering possibly into illegal circulation and program fluxion and the resulting that the system does not work in normal working order, auto-reset circuit is introduced, and then the system is ensured in normal running state. Meanwhile, reliability of system could be enhanced by used software design "trap" program. If microprocessor system were in trouble, CPU of microprocessor chip would visit auto-reset circuit at the stated time. And auto-reset circuit would be restarted and malfunction of the system is eliminated.

## 4 Experiments and Results

The experiments are done in the laboratory. Fire detector is fixed to the ceiling. Smoldering fire and flaming fire are produced by different material in turn, such as timber, cotton. Fire signal given off form the fire detector is comparatively in time. Time-signal curves are drawn as the following Fig.4 and Fig.5 shown. In the experiments, the time of given signal is related directly to distance form position of fire to location of fire detector. And then in the smoldering fire and flaming fire, time of signal given is not the same. In practice, smog particle is detected mainly in the smoldering fire. And temperature is detected mainly in the flaming fire. Because of hear diffused function, response time of temperature sensor lag to some extent. The proposed fire detector is detected fire comparative perfect. It has been observed from the graph ref Fig 6 distance between the hot object vs threshold exceedence time by sensor output voltage which shows the nature of the two heat sources at the same temperature but variable threshold excedence time. This helps estimates the possibility of fire at an early stage.

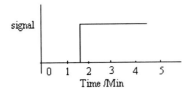

**Fig. 4.** Time-signal curves of Smoldering fire

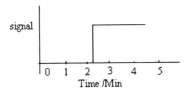

**Fig. 5.** Time-signal curves of flaming fire

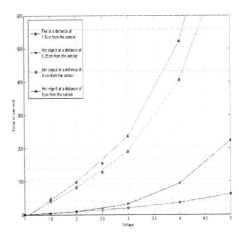

**Fig. 6.** Voltage Vs Threshold time graph

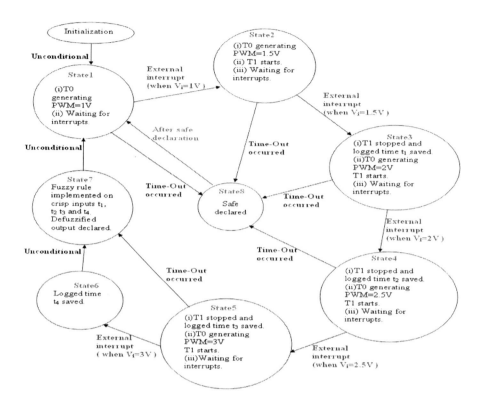

**Fig. 7.** State model of The Software

**Table 1.** The Simulated Fuzzy Outputs and Corresponding Result

| Sl. No | Source | Room Temperature | Temperature at The Surface of The Sensor | Distance | Output (LED Color & Alarm Status) |
|---|---|---|---|---|---|
| 1 | Hot Object | 26°C | 103°C | 0.5 cm | UNSAFE |
| 2 | Fire | 26°C | 107°C | 1.5 cm | FIRE |
| 3 | Hot Object | 26°C | 33°C | 4 cm | SAFE |
| 4 | Hot Object | 26°C | 21°C | 5 cm | SAFE |

## 5 Conclusions

Following experimentation is performed to identify the behavior of two different types of heat generating sources (More or less at the same temperatures) and take necessary steps to prevent the possible unwanted outcome of the event. Here the two heat generating sources are one is a hot object and other is fire. The temperature sensor placed at different distances and subsequent threshold exceedence time of the comparator through the processor has been recorded. This helps to understand the nature and characteristics at an early stage through sensor based reasoning techniques using fuzzy logic. The temperature of the environment is measured through a suitable measuring device under 4 different conditions and subsequent data has been logged. This real time data describes the threshold excedence time at 4 different stages like t1, t2, t3 and t4 as described above. The experimental data i.e. threshold accidences time and the subsequent difuzzified output has been shown in the Table.1 under 4 different conditions.

# References

[1] Nan, C.: Design of Fire Detection and Control Systems for Intelligent Buildings, vol. 7, pp. 19–60. Tsinghua University Press, Beijing (2001) (in Chinese)
[2] Tao, C., Hongyong, Y., Weicheng, F.: The developing fire detection technology. Fire Safety science 10(2), 108–112 (2001) (in Chinese)
[3] Brace, F.: Applying Advanced Fire-detection and Controls Technology to Enhance Elevator Safety During Fire Emergencies. Elevator World 44(4), 86–90 (1996)
[4] Shimanaihuatieseng, Er.: Fire warning device, vol. 8, pp. 19–120. Atomic Energy Press, Beijing (1995)
[5] Ran, H., Yuan, H.: An introduction of architecture fire safety engineering, vol. 11, pp. 15–148. University of Science and Technology of China Press, Hefei (1999) (in Chinese)
[6] Nebiker, P.W., Pleisch, R.E.: Photoacoustic gas detection for fire warning. Fire Safety Science 2, 173–180 (2001)
[7] Wahid, F., Givargis, T.: Embedded system design – A unified Hard ware / Software Introduction. Wiley, Chichester (2006)
[8] Patton, R., Frank, P., Clerk, R.: Fault diagnosis in dynamic systems – theory and application. Prentice Hall international series in systems and control engineering UK (1989)
[9] Calcutt, D., Cowhan, F., Parchizadeah, H.: 8051 microcontroller an application based introduction. Elsevier, Amsterdam (2004)
[10] Xiaohang, C.: Theory and Method of fire detected. Journal of china safety science 9, 24–29 (1999)
[11] Qizhong, L.: Principle and application of Single-chip Computerize, vol. 12, pp. 223–252. Beijing University of Aeronautics Astronautics Press, Beijing (2003)
[12] Konar, A.: Computational Intelligence: Principles, Techniques and Applications. Springer, Heidelberg (2005)

# A Scaled Test Bench for Vanets with RFID Signalling

Andrés Ortiz[*], Alberto Peinado, and Jorge Munilla

ETSI Telecomunicación, Dept. Ingeniería de Comunicaciones,
Universidad de Málaga Campus de Teatinos, 29071 Málaga, Spain
{aortiz,apeinado}@ic.uma.es

**Abstract.** Vehicular ad-hoc networks (vanets) are ad-hoc networks specially thought to provide communication among vehicles. The information in vanets can be shared among vehicles or sent in unicast, multicast or broadcast mode with the main goal of improving safety or even providing multimedia or commercial services. Several protocols and techniques have been proposed to be used in specific scenarios depending on the vehicle density and the spatial distribution. Anyway, it is necessary to have a tool for testing different scenarios and protocols proposed. Although simulation is usually a powerful technique for evaluating wireless or routing protocols, it is also necessary to have a real test bench to implement the protocols that could be already simulated with success. In this paper, we present a real framework for vanets that use wireless communication among vehicles and RFID for different traffic signalling information, such as bus stop or semaphore location, speed limit information, etc.

**Keywords:** Vanets, RFID, traffic safety, test bench.

## 1 Introduction

Vehicular networks interest is motivated by the capability of sending information among vehicles with the goal of improving the safety and comfort, as well as a range of new services specially thought for vehicles, including multimedia or commercial applications [4]. This has motivated an active research work in the last years, and having a specific tool for evaluating protocols or routing techniques developed for vehicular ad-hoc networks.

For evaluating the proposals we have two alternatives. The first consists of simulating vanets, using any simulation technique. In this sense, several applications have been developed in order to provide a specific tool for simulating the behaviour of vanets and simulation tools such as SUMO [2] or GrooveSim [4]. These tools provide a traffic model that could be used as an upper layer of a network simulator

---

[*] This work has been supported by Ministerio de Ciencia e Innovación of Spain and FEDER (EU) under grant TIN2008-02236/TSI (MUOVE: iMprovement of road safety throUgh planning, design and integration of cryptOgraphic services in VanEts).

such as NS2[1] or OPNET[5] for testing and evaluating network protocols. By the way, TraNS [6] provide a full simulation enviroment integrating SUMO, NS2 and Google Maps [7] as well as a specific interface between SUMO and NS-2. These tools are actually used for simulating vanets.

The second consists of testing the proposals using a demonstration framework. A demonstration framework implies to have scaled vehicles which includes the sensors, communication or movement parts we would have in a real vehicle.

In this paper we present such a demonstration framework including RFID for vehicle signalling. The main goal of this framework is to evaluate the security protocols defined and implemented by the MUOVE project [18], supported by the Spanish and European Government. Unlike other related projects, MUOVE is focused on the security of communication protocols employed for traffic safety or/and on-board comfort services. This fact determines the architecture of framework.

Thus, after this introduction, in Section 2 we describe the signalling method using passive RFID tags. In Section 3 we describe the behavioural model for the scaled vehicles since they could be remote controlled as well as autonomous (self-driven). In Section 4 we show the multiprocessor architecture of our model and the implementation details. Finally, Section 5 provides the conclusions of this work.

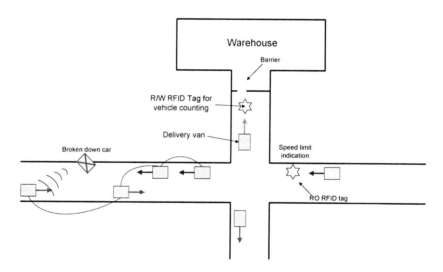

**Fig. 1.** A working vanet

## 2 Signalling Using RFID

In vanets we basically have two different sources of information. On the one hand, information coming from other vehicles, such as accident alerting or traffic jam indication [9]. On the other hand, information about the environment such as traffic signs, speed limits, motorway tools or any special semaphore. In both cases, communication has to provide a round trip journey for the information [9].

Since a characteristic in vanets is the fact of the base stations absence, communication among vehicles implies effective transmission and routing mechanisms for transferring the first type of information [10,11]. An alternative for the second type of information consist of having fixed nodes in any information point such as traffic signs, etc. However, this method implies to have a power source in any signalling point, which requires maintenance and therefore it is expensive. Other alternative consists of using passive RFID tags for signalling. Since the signal source is merely passive, it does not require any maintenance. Figure 1 shows a working vanet using RFID for signalling and vehicle counting applications. Thus, in this paper we present a scale model with programmable vehicles which also include wireless communication among them and a RFID infrastructure.

As previously commented, the main goal of having a RFID infrastructure for signalling is to avoid the need of power sources and maintenance. The RFID infrastructure consists of a RFID reader located at the bottom of the vehicle and a series of RFID tags located at any signalling point. So, the tags have to contain information regarding the point in which they are located (i.e.: a speed limit, a dangerous bend, etc.). Moreover, the tags could be moved as necessary.

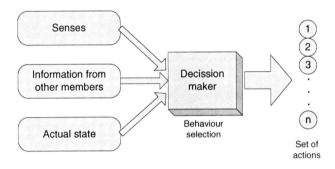

**Fig. 2.** Behavioural model

## 3 Behavioural Model

The purpose of the scale model is not only the implementation of already existing protocols and mechanisms for vanets but also the research and development of new applications. These applications go from new signalling methods to self-driving mechanisms. Thus, we need not only a programmable system but also architecture for implementing a behavioural model. In figure 2 we show the basic behavioural model implemented. With this model, the system processor acts as a decision maker, having into account the information coming from the sensors, from other members of the vanet, and its actual state (position, speed, etc.). This data is used as entry for the decision maker algorithm selecting a specific behaviour (i.e.: turn to the left, stop and wait, etc.).

In figure 3 we show a more detailed view of the behavioural model, in which, the sensors have a dynamic priority settled-up by the system processor though a feedback mechanism.

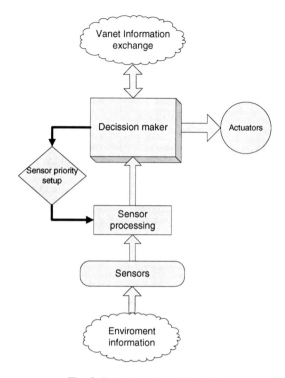

**Fig. 3.** Behavioural model detail

## 4 Scaled Vehicle Architecture

As shown in Figure 4, the scale model vehicle built for our demonstration framework contains the following elements: a main board which contains the system processor, the programmable wireless module and the motor drivers. The system processor is built with an atmega644 microcontroller [12] which provides up to 20 MIPS of processing power and 64 Kbytes of flash memory. This is enough for running a RTOS [13] real time operating system on it. This processor controls the vehicle and processes any incoming or outgoing information. A second board (sensor processing board) contains an atmega16 processor which receives information from the outside of the vehicle through the sensors, process it and send the result to the main processor which makes the behavioural decision. This architecture makes possible the parallel and cooperative work between the two processors, *offloading* the processing of the information coming from the outside of the vehicle on a CPU different from the system processor. So, we distribute the work between two processors, shifting part of the overhead to other processor different from the system processor. This technique provides a series of advantages as shown in [14].

By the way, implementing complex communication or routing protocols that uses artificial intelligence techniques [15] is possible with this versatile architecture that provides the necessary flexibility and reliability. Next we will see in detail the elements contained in each board.

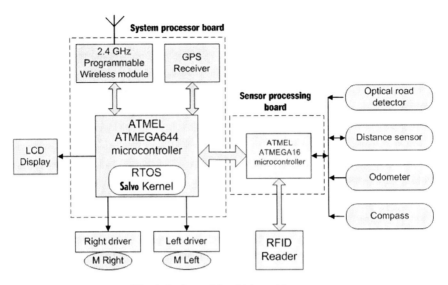

**Fig. 4.** Scale model vehicle architecture

### 4.1 System Board (Main Board)

The main board is composed to the following units:

**System processor:** atmega644 [12] microcontroller capable up to 20 MIPS with 64 Kbytes of flash memory. The Salvo [13] real-time operating system runs in this processor. This allows a real multitasking environment as well as a message-passing capability among tasks and a synchronization infrastructure using semaphores. This CPU processes the information coming from other vehicles and send information to the others using the wireless module. This way, the systems processor contains the secured communication and routing protocols, providing the necessary algorithms to ensure the correct and secure transference of the information to other vehicles as well as the behavioural code. This allows the implementation of self-driving algorithms.

The embedded real-time operating system also synchronizes the tasks runing on the sensor processor.

**Motor drivers:** consists of two power H-bridges for controlling the motors. H-Bridges provides the power to the motors for increasing or decreasing the velocity as well as breaking. The H-bridges are included on the main board.

**LCD Display:** which is a HD44780 based 4x16 LCD display, for showing any event occurrence. We use it for showing the actual state of the vehicle, including the behavioural mode the battery state, or any other useful variable for developing or debugging.

**Programmable wireless module:** consists of an autonomous programmable wireless module. This can be programmed just for transferring the information to other vehicles in a unicast, multicast or broadcast manner. This module does not only implement the physical layer but also the MAC and transport layers. It is also possible to implement an application layer for securing the communication using a 128-bit AES encryption algorithm. Nevertheless, in our architecture, the routing protocols are implemented on the system processor.

**GPS receiver:** Several routing algorithms use the absolute position information for routing the information packets [16, 17]. The GPS receiver provides the support for this kind of routing algorithms. The GPS receiver could be also used for self-driving aid.

### 4.2 Sensor Processing Board (Coprocessor Board)

The coprocessor board is composed to the sensor coprocessor and the communication interface with the sensors.

Sensor coprocessor is implemented on an atmel atmega16. This processor acts as an *offloading* engine for controlling the sensors. This processor reads the sensors and establishing a priority among them; nevertheless, making a behavioural decision. The sensors included on the scaled vehicle are classified in two categories: the first one for self-driving and the other for signalling and warning.

The sensors used for self-driving are:

*Distance detector.* This is implemented using a Sharp GP2D02 infrared ranger and an ultrasonic detector. This double sensor provides two capabilities: first, a radar-like system using the infrared detector, allowing the vehicle to know any object and the distance to it when running at low-speed. The second capability is a collision detector using an ultrasonic fixed range sensor.

*Odometer.* This sensor is used to know the run distance.

*Compass.* The compass allows to know the direction run. Used in with the odometer sensor, the vehicle knows the distance run and the direction. So, it is possible to compute the relative position of the vehicle.

*Optical road detector.* An optical sensor detects when the vehicle is out of the road. This causes the coprocessor send a high-priority notification to the system processor.

*RFID reader*/writer. As commented before, RFID technology is used in our vehicles for signalling. The RFID infrastructure consists of a reader included in each vehicle providing the capability of reading different nature signals from the outside. Thus, the vehicle identifies a tag on the road, which could be located on a traffic signal, crossroad, semaphore, etc. and definitely in every point which could be an event trigger on a vehicle. Nevertheless, tags located in special positions could be written by some vehicles. This could be useful for applications with entry barriers or delivery vehicles.

## 5 Conclusions

In this paper we describe a real test platform for vanets which is enough flexible to implement new routing algorithms, security protocols or self-driving mechanisms. The purpose of this scale model is to provide a useful platform for researching in security aspects of VANETs communications. Different VANETs aspects, actually under development, can also be tested in the platform. In this sense, we also have included a RFID signalling method with the goal to provide a range of new applications without the need of having a fixed infrastructure such as base stations network. On the other hand, the use of RFID in vanets give us the possibility of implement new applications that otherwise couldn't be implemented. Note that security in RFID is of vital importance in order to be applied on traffic safety systems.

# References

1. Network Simulator, http://nile.wpi.edu/NS/
2. SUMO. Simulation of Urban Movility, http://sumo.sourceforge.net
3. MOVE. Mobility model generator for Vehicular networks. School of Computer Science and Engineering. New South Wales. Australia, http://www.cse.unsw.edu.au/~klan/move
4. Mangharam, R., Weller, D.S., Stancil, D.D., Rajkumar, R., Parikh, J.S.: GrooveSim: A Topography-Accurate Simulator for Geographical Routing in Vehicular Networks
5. OPNET Network Simulator, http://www.opnet.com
6. Piorkowski, M., Raya, M., Lugo, A., Papadimitratos, P., Grossglauser, M., Hubaux, J.: TraNS: Realistic Joint Traffic and Network Simulator for VANETs. ACM SIGMOBILE Mobile Computing and Communications Review (2007)
7. Google Maps, http://maps.google.com
8. Eichler, S., Ostermaier, B., Schroth, C., Kosch, T.: Simulation of Car-to-Car Messaging: Analyzing the Impact on Road Traffic. Institute of Communication Networks, Munich University of Technology and BMW Research and Technology, Munich (2005)
9. Little, D.T.C., Agarwal, A.: An Information Propagation Scheme for Vanets. In: Proceedings of Intelligent Transportation Systems, September 2005, vol. (13-15), pp. 155–160 (2005)
10. Luo, J., Hubaux, J.P.: A Survey of Inter-Vehicle Communication. Tech. Rep., EPFL, Switzerland (2004)
11. Car to car communication consortium, "w3.car-to-car.org". OverDRiVE project, http://w3.ist-overdrive.org
12. http://www.atmel.com
13. Salvo Real Time Operating System for very low-cost embedded systems, http://www.pumpkinkinc.com
14. Ortiz, A., Ortega, J., Diaz, A.F., Cascón, P., Prieto, A.: Protocol Offload Analysis by Simulacion. Journal of Systems Architecture - Embedded Systems Design 55(1), 25–42 (2009)
15. Zhao, W., Liu, D., Jiang, Y.: Distributed Neural Network Routing Algorithm Based on Global Information of Wireless Sensor Network. In: International Conference on Communications and Mobile Computing (2009)
16. Stojmenovic, I.: Position-Based Routing in Ad Hoc Networks. IEEE Communications Magazine 40, 128–134 (2002)
17. Mauve, M., Widmer, J., Hartenstein, H.: A Survey on Position-Based Routing in Mobile Ad Hoc Networks. IEEE Network, 30–39 (November /December 2001)
18. MUOVE. Mejora de la segUridad vial mediante la planificación, diseño e integración de servicios criptOgráficos en VanEts, http://webpages.ull.es/users/cryptull/MUOVE/

# A SVM-Based Behavior Monitoring Algorithm towards Detection of Un-desired Events in Critical Infrastructures

Y. Jiang[1], J. Jiang[1], and P. Capodieci[2]

[1] Digital Media & Systems Research Institute, University of Bradford, UK
j.jiang1@bradford.ac.uk
[2] Selex Communications S.p.A, Italy
paolo.capodieci@Selex-Comms.com

**Abstract.** In this paper, we report our recent research activities under MICIE, a European project funded under Framework-7 Programme, in which a SVM-based behavior modeling and learning algorithm is described. The proposed algorithm further exploits the adapted learning capability in SVM by using statistics analysis and K-S test verification to introduce an automated parameter control mechanism, and hence the SVM learning and detection can be made adaptive to the statistics of the input data. Experiments on telecommunication network data sets support that the proposed algorithm is able to detect undesired events effectively, presenting a good potential for development of computer-aided monitoring software tools for protection of critical infrastructures.

## 1 Introduction

The term "Critical Infrastructures (CIs)" refers to "those assets or parts, which are essential for the maintenance of critical societal functions, including the supply chain, health, safety, security, economic or social well-being of people". According to the EU document COM(2006)787 [1], critical infrastructure encompass the following sectors and related sub sectors: (i) energy to include oil and gas production, refining, treatment, storage and distribution by pipelines, as well as electricity generation and transmission; (ii) information communication technologies (ICT) to include Internet, satellite, broadcasting, and instrumentation automation and control systems etc. (iii) water, food and health facilities and supply chains; (iv) transport and finance systems. As CIs can be damaged, destroyed or disrupted by deliberate acts of terrorism, natural disasters, negligence, accidents, computer hacking, criminal activities and malicious behaviours, it becomes extremely important to ensure that any disruptions or manipulations of CIs should be brief, infrequent, manageable, geographically isolated and minimally detrimental to the welfare of the member states, their citizens and the European Union. To this end, the MICIE project (funded under FP7) is to support the improvement of Critical Infrastructure Protection capability in Europe through the design and implementation of an on-line "MICIE alerting system", which is able to predict, in real time, the cascading effects (expressed in risk levels within the targeted QoS) on a given CI of some undesired events. It is expected that a range of software tools will be developed under MICIE via intelligent computing approaches to support the activities and decisions to be taken by the CI operators.

The challenging issue here is that critical infrastructures are dependent with each other in terms of their securities, and thus, to develop a useful MICIE alerting system, two categories of software tools need to be developed, which means that not only the behavior of each individual CI needs to be modeled, monitored and controlled, but also the inter-dependency among different CIs.

In this paper, we focus on the issue of behavior modeling of individual CIs, and hence a monitoring software tool can be developed first to detect unusual patterns and undesired events towards robust risk analysis, prediction and hence a range of meta-data can be produced to communicate with other CIs and pave the way for inter-dependency modeling as well as integrated approaches for the final MICIE alerting system.

Essentially, behavior modeling of individual CI is about analyzing the raw information generated to indicate the operational status of the CI. Examples of such raw information include the network traffic data for telecommunication CIs or sets of control parameters often referred to as KPIs (key performance indicators). Existing approach adopted by most of the CI industry is rule-based, where a set of thresholds is set up according to operational experiences, and anything beyond one or more thresholds could prompt investigation or other actions by engineers. While such approaches are effective to maintain a reasonable level of security, it is essentially labour intensive and thus the running cost is very high. Recently, artificial intelligence approaches are introduced to examine the possibility of computer-aided security of facilities to protect the CIs, especially the anomaly detection around the telecommunication networks [2-6]. These approaches are represented by statistics-based and neural network based. While statistics analysis [2,3], such as Bayesian and hidden Markov etc. is used to determine the most appropriate threshold values to complete the control and detection of those abnormal behaviors, the neural network based approaches [5,6] represents machine learning of training data set supported by ground truth to analyze the variation of input data and hence capture the behavior of the CI. In this paper, we combine these two approaches together to exploit the statistics analysis for adaptive estimation of controlling parameters to drive the machine learning approach, and hence an adaptive to input statistics machine learning approach can be designed for the purpose of modeling the behavior of CIs. In comparison with existing techniques, our proposed algorithm achieves a range of advantages, including: (i) the behavior modeling and unusual pattern detection are made adaptive to the variation of input data via its statistics analysis; (ii) automated determination of controlling parameters for SVM to achieve the best possible performances for pattern detection and modeling.

The rest of the paper is organized in two sections, where section 2 describes the proposed algorithm, section 3 reports experimental results and concluding remarks.

## 2 The Proposed Algorithm Design

As one of the most popular machine learning approaches, SVM has received tremendous attention in a number of areas to deal with learning, training and optimizing problems. General SVM uses a kernel-mapping technique to separate linearly non-separable cases in high dimensional space [8]. SVM not only separates data from different classes, but also separates data to its maximum margin. In other words, SVM not only divides mixed data sets into classes, but also optimizes such a classification. However, the weakness of general SVM lies in the fact that it requires labeled data sets to get it trained, yet many practical applications, especially the information associated with many CIs, do not have such labelled data to provide a so-called ground truth. This is because the operators do not have clear ideas about which patterns are regarded as abnormal and which are regarded as normal until investigation of individual cases is completed, and the outcome of such investigations is often case sensitive. To this end, we propose a statistics-based one-class SVM algorithm to complete the automated machine learning of CI's input information and hence leading to successful behavior modeling of CIs.

One-class-SVM is essentially an extension of support vector machines [7,9] used for detecting the outliers [8, 10]. The idea is that it first maps the data into high dimensional space, and then maximizes the margin between the mapped data and the origin. Apart from the fact that SVM is an unsupervised learning, the OCSVM also introduced a constraint parameter $V$ to control the maximum percentage of outliers in the dataset, which can be used to indicate a priori. As a result, the one-class-SVM is capable of being adaptive to input changes or different players when used to model the behavior of CIs. This is because that the priori specifies the maximum likelihood that outlier detections can be made, and hence such detection is less sensitive to changes of inputs generated by different CIs or different elements within a CI. In comparison with neural network based approaches [4], SVM presents a range of advantages, which can be summarized as: (i) while artificial neural networks (ANNs) can suffer from multiple local minima, the solution to an SVM is often global and unique; (ii) while ANNs use empirical risk minimization, SVMs use structural risk minimization. As a result, the computational complexity of SVM is not dependent on the dimensionality of the input space, and SVMs can also have a simple geometric interpretation and generate a sparse solution; (iii) from wide range of reports on evaluation of both SVMs and ANNs [10,11], it is generally concluded that SVMs often outperform ANNs in many classification-relevant applications due to the fact that they are less prone to over-fittings.

Given a data set describing the operational status of the target CI: $T = \{x_1, x_2, \cdots, x_l\}$, where $x \in R^N$ is an input vector with $N$ elements, i.e. each item inside the status data set is regarded as an N-dimensional vector, a learning system such as SVM-based can be established to process the N-dimensional vectors to model, analyze, and monitor the CI's operation. The essential task is to find a function $f$ that generates the value "+1" for most of the vectors in the data set, and "-1" for the other very small part. The Strategy for such a classification and detection is to use a one-class-SVM [10] and map the input data into a Hilbert space $H$ according to a mapping function $X = \phi(x)$, and separate the data from the origin to its maximum margin.

As a result, to separate the mapped data from the origin to its maximum margin is equivalent to solving the following quadratic optimization problem:

$$\min_{w \in F}: \tfrac{1}{2}\|w\|^2 + \frac{1}{vl}\sum_i \xi_i - \rho \qquad (1)$$

Subject to:

$$f(x) = w\phi(x_i) - \rho \geq -\xi_i, \xi_i > 0, i = 1,\cdots,l \qquad (2)$$

Where $v \in (0,1)$ is a constraint parameter to limit the maximum proportion of the high performance responses among all the ordinary responses as such that a maximum of $v \times 100\%$ are expected to return negative values according to $f(x) = w.\phi(x) - \rho$. $\xi_i$ are slack variables acting as penalties in the objective function.

It is proved [10] that $v \times 100$ is the upper bound percentage of the data that are expected to be outliers in the training data, and a vector $x_i$ is detected to be outlier in the training set, if and only if $\alpha_i = 1/(vl)$. $\alpha_i$ is the parameter directly determines the sensitivity of outlier detection using one-class-SVM. Its selection is dependent on the specifications and requirements for protection of individual CIs and specific expectations by the engineers who operate the CI protection. Inevitably, such parameter setting would be extremely complicated as it is connected to many other factors, such as investigation of suspicious events or patterns, their related human labour costs, and understanding of all the operational status information sets etc. As a matter of fact, such information data sets are normally provided by user partners within the MICIE consortium and hence making it difficult and time consuming for technology or research partners to understand such operational data sets before any artificial intelligence and machine learning algorithms could be developed. Further, CI operators are often sensitive in handing out critical information for confidential purposes, which make it additionally difficult to get the collaboration going smoothly.

Under this circumstance, we propose to adopt a light-touch approach, where focus of research is to analyze the operational data sets by estimating their statistics to provide a context simulation for machine learning without too much regards to their specific meaning to those CI operators. Otherwise, we could be trapped into the situation that research partners need to learn the whole CI operation process before any software tools could be developed, yet such learning process is often made extremely difficult and almost impossible for security reasons. In this way, research under MICIE becomes two important steps: (i) pre-processing the operational data sets and convert them into input vectors; (ii) estimate and analyze their statistics to activate the machine learning such as the one-class SVM as described above. In this paper, we report our initial results of our work in the second step, which is designed to solve the essential problem that how these parameters could be adaptively determined to drive the SVM.

Given the input operational data sets, $\psi = \{y_1, y_2, \cdots, y_M\}$, we estimate their statistics features, such as the mean and variance, as follows:

$$\mu = \frac{1}{M}\sum_{i=1}^{M} y_i \qquad (3)$$

$$\sigma^2 = \frac{1}{M}\sum_i (y_i - \mu)^2 \qquad (4)$$

We then apply K-S test to all the samples to verify a probability function distribution (PDF) to characterize the input information source, for which it is most likely expected to use Gaussian distribution. Consequently, we propose the following technique to estimate the SVM parameter $\gamma$ via exploiting the above statistics analysis of the input operational data sources:

$$\gamma = P(y_i > \mu + \sigma) \tag{5}$$

In other words, the constraint parameter is estimated to be the probability of those data samples that are larger than $\mu + \sigma$. As $\mu$ indicates the mean value of all samples and $\sigma$ the standard deviation, their combination would provide an adaptive constraint to characterize the input data source, which is equivalent to a threshold that is adaptive to the variation of input statistics and such adaptability is fully automatic. Figure-1 illustrates such an adaptability, where the shaded area indicates $P(y_i > \mu + \sigma)$. As seen, when the statistics of input changes, indicated by the change of $\mu$ or $\sigma$, the shaded area will also change correspondingly.

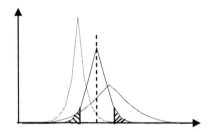

**Fig. 1.** Adaptability illustration to the variation of input statistics

## 3 Experimental Results and Concluding Remarks

To evaluate the proposed algorithm, we carried out two phases of experiments, where the first phase is to run the K-S test to verify the Gaussian distribution and the second phase is to detect unusual patterns out of telecommunication network traffic data sets, which is regarded as one of the most important CIs under the section of ICT.

Given the input samples, the K-S test is characterized by the following operation:

$$H = \begin{cases} 0 & \text{if } \gamma_{ks} = \max_{y \in \zeta_j, j \in [1,N]} \left| \sum_{i \leq y} C_i - \int f(y)dy \right| < D_\alpha \\ 1 & \text{else} \end{cases} \tag{6}$$

where $C_i$ is the ith bin value inside the differential data histogram.

The corresponding hypothesis is given as:

- H=0: It can be established that the tested samples come from a population with probability distribution f(y) at the given significance level $D_\alpha$;
- H=1: It can be established that the tested samples do not come from a population with probability distribution f(y) at the given significance level.

**Table 1.** K-S test results

| Sample sets | $f(y)$ | $\gamma_{KS}$ | $D_{0.05}$ | H |
|---|---|---|---|---|
| Set-1 | Gaussian | 0.0026 | 0.0059 | 1 |
|  | Laplace | 0.0265 | 0.0059 | 0 |
| Set-2 | Gaussian | 0.0023 | 0.0057 | 1 |
|  | Laplace | 0.0254 | 0.0057 | 0 |
| Set-3 | Gaussian | 0.0047 | 0.0060 | 1 |
|  | Laplace | 0.0234 | 0.0060 | 0 |

Table-1 illustrates all the K-S test results, where we divided the input samples into three sub-sets for efficiency of analysis purposes, and tested both Gaussian distribution and Laplace distribution. As seen, all the test results indicate that the samples we tested conform to the Gaussian distribution for a number of significance levels. In case that the K-S test indicates a strong fit-in with Laplace distribution, the constraint parameter should be determined via Laplace PDF in equation (5). As seen, the advantage of the proposed algorithm lies in the fact that the SVM learning is not only made adaptive to the statistics features of the input data, but also to the varying nature of its probability distributions. For complicated behavior of CIs, such approach will prove useful that a number of PDFs are required to present a piece-wise simulation and characterization of the input information source.

Figure-2 illustrates the experimental results of the outlier detection via the one-class SVM learning mechanism for two different sets of traffic data captured from two network nodes.

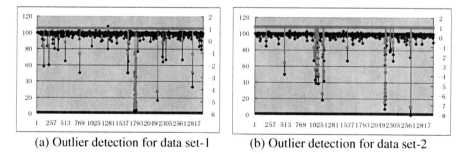

(a) Outlier detection for data set-1     (b) Outlier detection for data set-2

**Fig. 2.** Illustration of experimental results for outlier (undesired events) detection

While the blue points represent the normal behavior of the telecommunication network nodes, the pink highlighted points indicate suspicious patterns or events, which may need attention or even investigations by the CI operators. Via our industrial parameter's investigations, it is verified that the proposed algorithm achieved around 80% accuracy in detecting those true undesired events in comparison with the ground truth. The associated false positive rate is around 17%, which is a reasonable price to be paid for the outlier detection. There exist certain relationship between the false positive rate and true positive rate, and the general trend is that the closer the true positive rate is to 100%, the larger the false positive rate, which could reach 100% in certain cases.

In this paper, we have described our latest research activities under the MICIE project, where a SVM-based machine learning algorithm is proposed to automatically monitor the behavior of CIs and detect unusual patterns or events out of the operational data sets. The essential idea is to carry out statistics analysis to pre-process the input information and hence provide a set of context information to drive the machine learning module and make it adaptive to the statistics of input. In this way, the machine learning algorithm can be made sort of universal, in which the meaning of input data could be less regarded via concentrating on capturing their statistics rather than their functionalities. Experimental results support that such an approach is effective and also efficient in terms of running costs, due to the fact that the entire process is computer-aided, self-adapted, and the exposed parameters are tractable by human users. In practical cases, the statistics estimation and analysis need to be regularly run to ensure that the hypothesis of a certain probability distribution is correct, and the specific values of mean and standard deviation are updated.

Finally, the authors wish to acknowledge the financial support for the research work supported by the MICIE project under the European Framework-7 Programme (Contract No: 225353).

## References

1. http://www.cordis.lu/ist/projects/projects.htm
2. Patcha, A., Park, J.-M.: An overview of anomaly detection techniques: Existing solutions and latest technological trends. Computer Networks 51, 3448–3470 (2007)
3. Rajasegarar, S., Leckie, C., Palaniswami, M.: Anomaly detection in wireless sen-sor networks. IEEE Wireless Communications (August 2008)
4. Han, S.-J., Cho, S.-B.: Evolutionary Neural Networks for Anomaly Detection Based on the Behavior of a Program. IEEE Transactions on Systems, Man, and Cybernetics, Part B: Cybernetics 36(3) (2006)
5. Muthuraman, S., Jiang, J.: Anomaly detection in telecommunication network performance data. In: Proceedings of the 2007 International Conference on Artificial Intelligence, Monte Carlo Resort, Las Vegas, Nevada, USA (June 2007)
6. Shon, T., Moon, J.: A hybrid machine learning approach to network anomaly detection. Information Sciences 177, 3799–3821 (2007)
7. Manevitz, L.M., Yousef, M.: One-Class SVMs for document classification. Journal of Machine Learning Research 2, 139–154 (2001)
8. Keerthi, S.S., Lin, C.J.: Asymptotic behaviors of support vector machines with Gaussian Kernel. Neural Computation 15(7), 1667–1689 (2003)
9. Schölkopf, B., Williamson, R., et al.: Support vector method for novelty detection. In: Neural Information processing Systems, pp. 582–588. MIT Press, Cambridge (2000)

10. Li, Y., Jiang, J.: Combination of SVM knowledge for microcalcification de-tection in digital mammograms. LNCS, vol. 317, pp. 359–365. Springer, Heidelberg (2004)
11. Kalatzis, I., Piliouras, N., et al.: Comparative evaluation of probabilistic neural network versus support vector machines classifiers in discriminating EPR signals of depressive patients from healthy control. In: Image and Signal Processing and Analysis, ISPA 2003, September 18-20, vol. 2, pp. 981–985 (2003)

# Design and Implementation of High Performance Viterbi Decoder for Mobile Communication Data Security

T. Menakadevi and M. Madheswaran

Anna University Coimbatore, Coimbatore, Tamil Nadu, India
menaka_sar@rediffmail.com, madheswaran.dr@gmail.com

**Abstract.** With the ever increasing growth of data communication in the field of e-commerce transactions and mobile communication data security has gained utmost importance. However the conflicting requirements of power, area and throughput of such applications make hardware cryptography an ideal choice. Dedicated hardware devices such as FPGAs can run encryption routines concurrently with the host computer which runs other applications. The use of error correcting code has proven to be are effective way to overcome data corruption in digital communication channel. In this paper, we describe the design and implementation of a reduced complexity decode approach along with minimum power dissipation FPGAs for Mobile Communication data security.

**Keywords:** Data security, FPGA, Hardware Cryptography, Viterbi algorithm.

## 1 Introduction

Convolutional encoding with Viterbi decoding is a powerful method for forward error correction in Mobile communication data security applications. It has been widely deployed in many communication data security systems to improve the limited capacity of the communication channels. With the proliferation of portable and mobile devices as well as the ever increasing demands for high speed data transmission, both power minimization and high speed with reduced hardware complexity have become important performance metrics when implementing Viterbi decoders.

The rapid evolution of modern FPGAs have led to the inclusion of up to multi-million gate configurable logic and various precompiled hardware components (e.g. memory blocks, dedicated multipliers, etc.) on a single chip. Examples of these FPGA devices are Xilinx Spartan-3/Virtex- II/Virtex-II Pro [2] and Altera Stratix/Stratix-II [1]. These FPGA devices provide high computational performance, low power dissipation per computation, and reconfigurability.

### 1.1 An Overview

Several researchers have implemented Viterbi decoders on FPGAs [4], [5] (See Section 2 for more details). However, these designs use a lot of registers and multiplexers and thus consume a lot of power. In this paper, we propose a circular linear pipeline architecture based on the trace-back algorithm for implementing Viterbi decoders on FPGAs.

The two major advantages provided by our design are: (1) *area efficient*: The Viterbi algorithm, which is the most extensively employed decoding algorithm for convolutional codes, works well for less-complex codes, indicated by *constraint length K*. However, the algorithm's memory requirement and computation count pose a performance obstacle when decoding more powerful codes with large constraint lengths. In order to overcome this problem, the Adaptive Viterbi algorithm (AVA) [4] [6] has been developed.

This algorithm reduced the average number of computations required per bit of decoded information while achieving comparable bit error rates (BER) versus Viterbi algorithm implementations. The tracing back, updating, and storing of the input information sequences are accomplished concurrently within the Configurable Logic Blocks(CLBs) that constitute the circular linear pipeline. Our circular linear pipeline architecture overcomes the low throughput problem in the previous implementations of the trace-back algorithm. The degree of parallelism of our design is parameterized and is determined by the number of CLBs employed; (2) *minimum power dissipation*. First, by employing a adaptive viterbi decoder with trace-back algorithm, our design greatly reduces data movement compared with the register-exchange algorithm. Switching activity, a dominating factor that affects power dissipation, is also reduced.

Second, embedded memory blocks are used as the main storage. Embedded memory blocks dissipate much less power per bit data than that of slice-based memory blocks and flip-flop based registers. The paper is organized as follows. Section 2 discusses Viterbi Decoding Algorithm. Our implementation of the Viterbi decoding algorithm is presented in Section 3. Experimental results are shown in Section 4. We conclude in Section 5.

## 2 Viterbi Decoding Algorithm

Viterbi algorithm can be classified into two categories, register exchange and trace-back, depending on how the information of the surviving paths is stored.

The *register-exchange* algorithm stores the surviving paths at each step of the decoding trellis directly. When the data comes in, each pair of surviving paths compete with each other in the next step of the decoding trellis. The complete surviving path is copied to the place where the discarded path is stored. Since the surviving paths are stored explicitly, when the truncation length is reached, the output symbol can be obtained immediately from the end of the surviving path with the greatest possibility of matching the input data sequence.

An implementation of the register exchange algorithm on FPGAs is proposed by Swaminathan *et. al.* [4]. They realized a suboptimal Viterbi decoding algorithm. In their design, copying of the surviving paths is realized using multiplexers and registers. The high memory bandwidth required by the concurrent copying of surviving paths prevents the use of embedded memory blocks even though embedded memory blocks are much more power efficient than registers for storage. This is because access of the data stored at the memory blocks is restricted to two 32-pin I/O ports, which cannot provide such a high memory bandwidth. Another drawback is that all the multiplexers and registers are active during the decoding procedure, which consume a lot of power. Hence, while their design provides high throughput, it is not suitable for power constrained systems.

In contrast, using techniques such as the one-bit function proposed in [6], the *trace-back* algorithm stores the transitions between two consecutive steps on the trellis. No path copying is required when a path is discarded. Since the path information is stored implicitly, when the truncation length is reached, a trace-back stage is required to trace back along the path transitions in order to identify the output symbol. An implementation of the trace-back algorithm can be found in a low-power architecture for Viterbi decoding proposed by Liang *et. al.* [7].

Embedded memory blocks are used to improve power efficiency. For environments with low SNR, a long truncation length is required and their decoder results in a very low data throughput. Xilinx provides two implementations of Viterbi decoders on their FPGA devices [2]. The parallel implementation employs the register-exchange algorithm while the serial implementation employs the trace-back algorithm. Both versions suffer from either high power dissipation or low throughput problem as described above.

Truong *et. al.* [5] propose a fully pipelined VLSI architecture to implement Viterbi decoding using a trace-back algorithm. Their design achieves a high throughput of one output symbol per clock cycle. However, in their architecture, all the storage registers are active during the decoding process. This would result in a high memory I/O bandwidth requirement. Such requirement cannot be sustained by the embedded memory blocks on FPGAs. It would also result in high power dissipation due to significantly high switching activity. Therefore, the design in [5] is not suitable for implementation on FPGAs .

## 3 Our Design

Our design of Viterbi decoder is based on the trace-back algorithm. The overall architecture of our design is shown in Figure 1. It consists of two major components: the branch selection unit and the trace-back unit. The *branch selection unit* calculates the partial costs of the paths traversing the decoding trellis, selects surviving paths, and identifies partial paths with lowest costs. The memory blocks within the *trace-back unit* store the selection results from the branch selection unit. Using the stored path selection information, the *trace-back unit* traces back along the paths with lowest costs identified by the branch selection unit and generates output symbols. We employ a circular linear pipeline of PEs and move the data along the linear pipeline so that the trace-back operations can be performed concurrently within the PEs. Details of these two components are discussed in the following subsections.

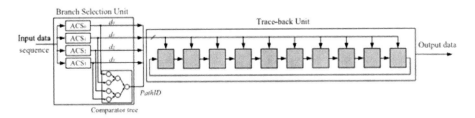

**Fig. 1.** Architecture of the proposed Viterbi decoder

## 3.1 Branch Selection Unit

Let $K$ denote the constraint length of the convolutional code and $TL$ denote the truncation length of Viterbi decoder. Then, there are $2K-1$ surviving paths at each step of the decoding trellis and one output symbol is generated after tracing back $TL$ steps on the trellis.

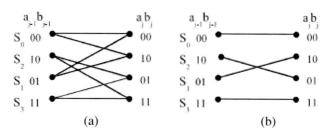

**Fig. 2.** (a) Possible connections between two consecutive steps on the trellis; (b) one selection path result for $(d_0, d_1, d_2, d_3) = (0, 1, 0, 1)$

Since the branch selection unit is relatively simple, we fully parallelize its operations in order to achieve a high data throughput. The branch selection unit is clock gated so that it dissipates negligible amount of power when there is no input data. There are $2K-1$ *Add-Compare-Select* units, ACS k, $0 \leq k \leq 2K-1 - 1$. Each ACS unit is responsible for selecting one of the $2K-1$ surviving paths and calculating its cost pCst. The path selection results, dk, $0 \leq k \leq 2K-1 - 1$, is represented using the one-bit function proposed in [5]. For example, we consider the possible connections between two consecutive steps on the decoding trellis as shown in Figure 2(a). For each pair of paths merging into a node on the right, only one of them is selected. dk = 0 if the path coming from above is selected; dk = 1 if the path coming from below is selected. The selection result shown in Figure 2(b) is represented as (d0, d1, d2, d3) = (0, 1, 0, 1). Finally, the costs of the surviving paths are sent to the comparator tree where the path with the lowest cost is identified. The comparator tree is fully-pipelined and is composed of $2K - 1$ comparator modules. The architecture of these comparator modules is shown in Figure 3. Each module accepts the partial path IDs (pID1 and pID2) and the costs (pCst1 and pCst2) of the two input paths and outputs the partial path ID (pID3) and cost (pCst3) of the path with the lower cost.

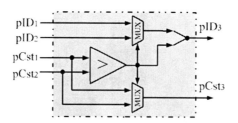

**Fig. 3.** Comparator Module

## 3.2 Trace-Back Unit

As shown in Figure 1, the trace-back unit consists of a circular linear pipeline of Np processing elements (PEs), PEi, $0 \leq i \leq Np - 1$. Np is a user-defined parameter and determines the throughput of the Viterbi decoder.

The architecture of a PE is shown in Figure 4. The path selection information from the branch selection unit is stored at the survivor memory. Instead of the centralized architecture adopted by the previous designs, we implement the survivor memory in a distributed manner. Each PE has its own survivormemory implemented using the embedded memory blocks (BRAMs) on Xilinx FPGAs. These BRAMs provide two I/O ports (port A and port B) that enable independent shared access to a single memory space.

The data from the branch selection unit comes into the survivor memory through port A of the BRAMs and is made available to the trace-back circuits through port B. Let L denote the trace-back depth of the PEs. The memory space of each PE is divided into two parts, each of which stores the trace-back information of L consecutive steps on the decoding trellis. These two parts of memory space are used alternatively to provide trace-back information to the trace-back circuit through port B and to store the data from the branch selection unit through port A.

The lower part of Figure 4 shows the trace-back circuit when $K = 3$. yi,j(ai,jbi,j) are the path selection information from the survivor memory. Regs are registers that store (ai,jbi,j), the status of trace-back path i at step j. According to the definition of one-bit function, the state of trace-back path i at step $j - 1$ can be obtained using the following equations.

$$ai,j - 1 = bi,j \tag{1}$$

$$bi,j - 1 = yi,j(ai,jbi,j) \tag{2}$$

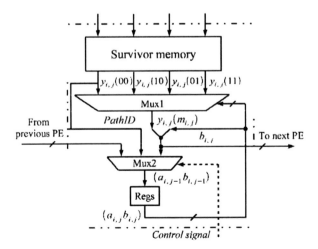

**Fig. 4.** P E$_i$ in the circular linear pipeline

Equation 2 is implemented using multiplexer MUX1 in Figure 4. The data required by the concurrent trace-back processes is stored in a distributed manner among the PEs. Multiplexer MUX2 is responsible for moving the trace-back processes along the circular linear pipeline so that these processes can be executed in the PEs where the required trace-back data is stored.

## 4 Experimental Approach

To allow for parameter testing for the High Performance Viterbi algorithm, a sample data transmission system was first modeled in software. The overall communication system model that has been used for experimentation is shown in Figure 5. This system contains blocks for data generation, encoding, transmission, and decoding.

The *Random Bit Generator* is a *C* module that generates a randomized bit sequence to model transmitted data. The *convolutional encoder* can be parameterized to assorted constraint lengths. The encoded data from the encoder is fed to the *AWGN channel simulator*. This block simulates a noisy channel where errors are inserted into the bit sequence. The amount of noise depends on the Signal-to-Noise-Ratio (*SNR*) preset by the user. The channel is noisy if *SNR* is low. The symbols obtained from the AWGN channel model are quantized before being sent to the *decoder* as its input. On receiving the input, the decoder attempts to recover the original sequence.

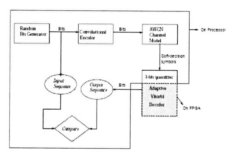

**Fig. 5.** System Model

For hardware experimentation, the High Performance Viterbi algorithm design was mapped to a Xilinx Spartan-III FPGA which is part of a FPGA board. This allowed for in-field testing of VA designs for constraint lengths up to $K=9$ with the rest of the communication system modeled in software. An RTL level description of the adaptive Viterbi decoder was written in VHDL that could be mapped to Spartan III device. The VHDL code was simulated using Mentor Graphics Model Sim tools. All designs were synthesized, simulated and mapped to Xilinx hardware using Xilinx foundation series ISE 9.1i tools with timing constraints. The maximum frequencies of operation of the FPGAs were obtained from the Xilinx *TRACE* timing analyzer tool.

The logic resources used by the High performance Viterbi decoder architecture was measured in terms of logic block (CLB) usage. Table 1 summarizes the resource utilization of the High Performance Viterbi decoder for different constraint lengths from $K = 4$ to $K = 9$ on the Xilinx Spartan III FPGA. As shown in Table 1, the proposed Viterbi decoder of constraint length 9 utilized 100% of 1296 Spartan III CLB resources (85% LUT utilization).

The power values were obtained by using XPower to analyze the simulation files from ModelSim which record the switching activities of each logic and wire on the device. The throughput of our Viterbi decoders increases linearly with the number of CLBs, Np.

The power dissipation for decoding one bit data also increases due to the costs for sending input data to each CLB and moving the trace-back processes along the linear pipeline. When Np = 1, our designs lead to the serial implementations of the trace-back algorithm discussed in Section 2. Thus, designs with Np = 1 are used as the baseline to illustrate the performance improvement. We consider throughput, power and complexity as the performance metric. Designs with Np = 32 achieve maximum performance improvements of 26.1% when K = 6 and 18.6% when K = 8 compared with the baseline designs with Np = 1.

**Table 1.** FPGA resource utilization for the adaptive Viterbi decoder K=(4 to 9)

| K | Nmax | T | TL | CLBs | 4-LUTs | 3-LUTs |
|---|------|----|----|------|--------|--------|
| 4 | 4 | 14 | 20 | 548 | 948 | 204 |
| 5 | 7 | 14 | 25 | 1163 | 1198 | 356 |
| 6 | 8 | 18 | 30 | 1196 | 2077 | 496 |
| 7 | 8 | 17 | 35 | 1207 | 2096 | 543 |
| 8 | 8 | 17 | 40 | 1265 | 2103 | 586 |
| 9 | 9 | 18 | 45 | 1276 | 2196 | 640 |

Moreover, there are more advantages of this proposed design on having smaller slices and block RAMs and operating with adaptive coding rates of 1/2, 1/3, 1/4 and 1/6 as well. Finally, it has been tested in an AWGN channel, the BER (bit error rate) of this work for rates 1/2, 1/3, 1/4 and 1/6 are performed and shown in Figure 6.

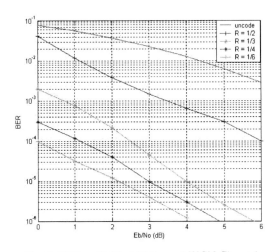

**Fig. 6.** BER vs Eb/No (dB) in an AWGN Channel

## 5 Conclusion

An architecture for implementing high Performance Viterbi decoders on FPGAs for Mobile communication data security is proposed. The dedicated hardware device such as FPGAs can run encryption routines concurrently with minimum power dissipation. We measured the dynamic power dissipation of a viterbi decoder circuit based on the annotated switching activity from gate level simulation. The effectiveness of our design is shown through the analysis of the architecture as well as low-level experimental results. The use of multiple clock domains as in the Xilinx design of Viterbi decoder can also be applied to our design to further improve speed and minimum power dissipation. The proposed architecture is well suited for Mobile Communication data security applications, with minimum power dissipation. So that, the dedicated hardware cryptography can run at high speed with maximum throughput.

## References

1. Altera, Inc., http://www.altera.com
2. Xilinx, Inc., http://www.xilinx.com V - 36
3. Dick, C.: The Platform FPGA: Enabling the Software Radio. Software Defined Radio Tech. Conf. SDR (2002)
4. Swaminathan, S., Tessier, R., Goeckel, D., Burleson, W.: A Dynamically Reconfigurable Adaptive Viterbi Decoder. ACM FPGA (2002)
5. Truong, T., Shih, M.-T., Reed, I.S., Satorius, E.: A VLSI Design for A Trace-back Viterbi Decoder. IEEE Trans.Comm (March 1992)
6. Forney, G.: The Viterbi Algorithm. Proc. IEEE (1973)
7. Liang, J.: Development and Verification of System-On-a- Chip Communication Architecture. Ph.D. Thesis, Univ. of Mass (2004)
8. Mentor Graphics, Inc., http://www.mentor.com
9. Tuttlebee, W.: Software Defined Radio: Enabling Technology. John Wiley & Sons, Chichester (2002)
10. Mukhopadhyay, D., RoyChowdhury, D.: An Efficient End to End Design of Rijndael Cryptosystem in 0.18 $\mu$ m CMOS. In: Proceedings of the 18th International Conference on VLSI Design 2005 jointly held with 4th International Conference on Embedded Systems Design, Kolkata, India (2005)

# An Adaptive Multi-agent Solution to Detect DoS Attack in SOAP Messages

Cristian I. Pinzón, Juan F. De Paz, Javier Bajo, and Juan M. Corchado

Departamento Informática y Automática, Universidad de Salamanca,
Plaza de la Merced s/n 37008, Salamanca, Spain
{cristian_ivanp,fcofds,jbajope,corchado}@usal.es

**Abstract.** A SOAP message can be affected by a DoS attack if the incoming message has been either created or modified maliciously. The specifications of existing security standards do not focus on this type of attack. This article presents a novel distributed and adaptive approach for dealing with DoS attacks in Web Service environments, which represents an alternative to the existing centralized solutions. The solution proposes a distributed hierarchical multi-agent architecture that implements a classification mechanism in two phases. The main benefits of the approach are the distributed capabilities of the multi-agent systems and the self-adaption ability to the changes that occur in the patterns of attack. A prototype of the architecture was developed and the results obtained are presented in this study.

**Keywords:** Multi-agent System, CBR, Web Service, SOAP Message, DoS attacks.

## 1 Introduction

The Web services processing model requires the ability to secure SOAP messages and XML documents as they are forwarded along potentially long and complex chains of consumer, provider, and intermediary services. However all standards that have been proposed to date, such as WS-Security [1], WS-Policy [2], WS-Trust [3], WS-SecureConversation [4], etc. focus on the aspects of message integrity and confidentiality and user authentication and authorization [5].

Until now, denial-of-service (DoS) attacks have not been dealt with in Web Services environments. A DoS attack on Web Services takes advantage of the time involved in processing XML formatted SOAP messages. The DoS attack is successfully carried out when it manages to severely compromise legitimate user access to services and resources. XML messages must be parsed in the server, which opens the possibility of an attack if the messages themselves are not well structured or if they include some type of malicious code. Resources available in the server (memory and CPU cycles) can be drastically reduced or exhausted while a malicious SOAP message is being parsed.

This article presents a novel distributed multi-agent architecture for dealing with DoS attacks in Web Services environments. Additionally, the architecture has a four-tiered hierarchical design that is better capable of task distribution and error recovery. The most important characteristic of the proposed solution is the two-phased mechanism that was designed to classify SOAP messages. The first phase applies the initial filter for detecting simple attacks without requiring an excessive amount of resources. The second phase involves a more complex process which ends up using a significantly higher amount of resources. Each of the phases incorporates a CBR-BDI [6] agent with reasoning, learning and adaptation capabilities. The CBR engine initiates what is known as the CBR cycle, which is comprised of 4 phases. The first agent uses a decision tree and the second a neural network, each of which is incorporated into the respective reuse phase of the CBR cycle. As a result, the system can learn and adapt to the attacks and the changes in the techniques used in the attacks.

The rest of the paper is structured as follows: section 2 presents the problem that has prompted most of this research. Section 3 focuses on the design of the proposed architecture. Finally, section 4 presents the results and conclusions obtained by the research.

## 2 DoS Attacks Description

One of the most frequent techniques of a DoS attack is to exhaust available resources (memory, CPU cycles, and bandwidth) on the host server. The probability of a DoS attack increases with applications providing Web Services because of their intrinsic use of the XML standard. In order to obtain interoperability between platforms, communication between web servers is carried out via an exchange of messages. These messages, referred to as SOAP messages, are based on XML standard and are primarily exchanged using HTTP (Hyper Text Transfer Protocol) [7]. The server uses a parser, such as DOM, Xerces, etc. to syntactically analyze all incoming XML formatted SOAP messages. When the server draws too much of its available resources to parse SOAP messages that are either poorly written or include a malicious code, it risks becoming completely blocked.

Attacks usually occur when the SOAP message either comes from a malicious user or is intercepted during its transmission by a malicious node that introduces different kinds of attacks.

The following list contains descriptions of some known types of attacks that can result in a DoS attack, as noted in [8, 9, 10].

- **Oversize Payload:** It reduces or eliminates the availability of a Web Service when the CPU, memory or bandwidth are being tied up by the processing of messages with an enormous payload.
- **Coercive Parsing:** Just like a message written with XML, an XML parser can analyze a complex format and lead to an attack when the memory and processing resources of the server are being used up.
- **Injection XML:** This is based on the ability to modify the structure of an XML document when an unfiltered user entry goes directly to the XML stream. The message is captured and modified during its transmission.

- **SOAP header attack:** Some SOAP message headers are overwritten while they are passing through different nodes before arriving at their destination.
- **Replay Attack:** Sent messages are completely valid, but they are sent en masse over short periods of time in order to overload the Web Service server.

All Web Services security standards focus on strategies independent from DoS attacks. Other measures have been proposed that do not focus directly on dealing with DoS attacks, but can deal with some of the inherent techniques. One solution based on the XML firewall was proposed to protect Web Services in more detail [8]. By applying a syntactic analysis, a validation mechanism, and filtering policies, it is possible to identify attacks in individual or group messages. An adaptive framework for the prevention and detection of intrusions was presented in [9]. Based on a hybrid focus that combines agents, data mining and diffused logic, it is supposed to filter attacks that are either already known or new. The solution as presented is an incipient idea still being developed and implemented. Finally, another solution proposed the use of Honeypots as a highly flexible security tool [11]. The focus incorporates 3 components: data extraction based on honeyd and tedpdum data analysis, and extraction from attack signatures. Its main inconvenience is that it depends too much on the ability of the head of security to define when a signature is or is not a type of attack. Even when the techniques mentioned claim to prevent attacks on Web Services, few provide statistics on the rates of detection, false positives, false negatives and any negative effects on application performance.

## 3 An Agent Based Architecture

Agents are characterized by their autonomy; which gives them the ability to work independently in real-time environments [12]. Furthermore, when they are integrated within a multi-agent system they can also offer collaborative and distributed assistance in carrying out tasks [13].

The main characteristic of the multi-agent architecture proposed in this paper is the incorporation of CBR-BDI [6] deliberative agents. The CBR-BDI classifier agents implemented here use an intrusion detection technique known as anomaly detection. In order to carry out this type of detection, it is necessary to extract information from the HTTP/TCP-IP protocol headers that are used for transporting messages, as well as from the structure and content of the SOAP message, the message processing tasks, and any activity among Web Services users.

Our proposal is a distributed, hierarchical multi-agent architecture integrated for 4 levels with distinct BDI agents. The hierarchical structure makes it possible to distribute tasks on the different levels of the architectures and to define specific responsibilities. Meanwhile, it is essential to maintain a continuous interaction and communication between agents, as they must be able to continue requesting services and deliver results. The architecture presented in figure 1 shows the four levels with BDI agents organized according to their roles.

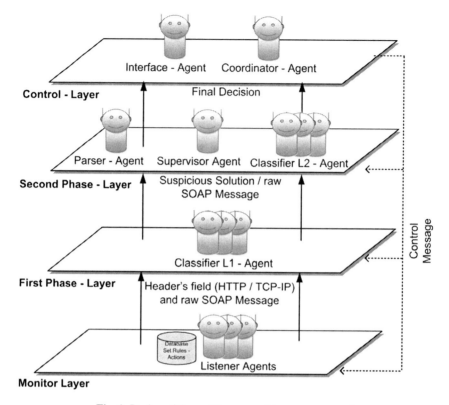

**Fig. 1.** Design of the multi-agent architecture proposed

The following section describes the functionality of the agents located in each layer of the proposed hierarchical structure.

- **Listener agents:** Capture any traffic directed towards the server and process the HTTP and TCP/IP protocol headers that are used to transport SOAP messages. JPCAP is used to identify and capture any traffic that contains SOAP message packets. The information extracted from the HTTP/TCP-IP transport protocol message headers is sent to the next layer in order to carry out the classification process. In addition to these tasks, Listener agents use an IP register to monitor user activities. This type of monitoring makes it possible to identify suspicious activities similar to message replication attacks.
- **Classifier-L1 agents:** These CBR-BDI agents are located on layer 2 of the architecture and are in charge of executing the first phase of the classification process based on the data sent by the Listener agents. These agents initiate a classification by incorporating a CBR engine that in turn incorporates a decision tree strategy in the reuse phase. The main goal of this initial phase is to carry out an effective classification, but without requiring an excessive amount of resources. The case description for this CBR is shown in Table 1 and involves various fields extracted from the HTTP/TCP-IP transport protocol message headers.

**Table 1.** Case Description First Phase - CBR

| Fields | Type |
| --- | --- |
| IDService | Int |
| Subnet mask | String |
| SizeMessage | Int |
| NTimeRouting | Int |
| LengthSOAPAction | Int |
| TFMessageSent | Int |

Within the CBR cycle, specifically in the reuse phase, a particular classification strategy is used by applying a knowledge extraction method known as Classification and Regression Tree (CART), which creates 3 groups: legal, malicious and suspicious. Messages that are classified as legal are sent to the corresponding Web Service for processing. Malicious messages are immediately rejected, while suspicious messages continue through to the classification process.

- **Classifier-L2 agents:** These CBR-BDI agents carry out the second phase of classification from layer 3 of the architecture. In order to initiate this phase, it is necessary to have previously started a syntactic analysis on the SOAP message to extract the required data. This syntactic analysis is performed by the Parser Agent. Once the data have been extracted from the message, a CBR mechanism is initiated by using a Multilayer Perceptron (MLP) neural network in the reuse phase. Table 2 presents the fields used in describing the case for the CBR in this layer.

**Table 2.** Case Description Second Phase - CBR

| Fields | Type |
| --- | --- |
| SizeMessage | Int |
| NTimeRouting | Int |
| LengthSOAPAction | Int |
| MustUnderstandTrue | Boolean |
| NumberHeaderBlock | Int |
| NElementsBody | Int |
| NestingDepthElements | Int |
| NXMLTagRepeated | Int |
| NLeafNodesBody | Int |
| NAttributesDeclared | Int |
| CPUTimeParsing | Int |
| SizeKbMemoryParser | Int |

The neural network is trained from the similar cases that were recovered in the retrieve phase. Once the neural network has been trained, the new case is presented to the network for classification as either legal or malicious. However, because there is a possibility of significant errors occurring in the training phase of the neural network, an expert will be in charge of the final review for classifying the message.

- **Supervisor Agent:** This agent supervises the Parser agent since there still exists the possibility of an attack during the syntactic processing of the SOAP message. This agent is located in layer 3 of the architecture.
- **Parser Agent:** This agent executes the syntactic analysis of the SOAP message. The analysis is performed using SAX as parser. Because SAX is an event driven API, it is most efficient primarily with regards to memory usage, and strong enough to deal with attack techniques. The data extracted from the syntactic analysis are sent to the Classifier L-2 agents. This agent is also located on layer 3 of the architecture.
- **Coordinator Agent:** This agent is in charge of supervising the correct overall functioning of the architecture. Additionally, it oversees the classification process in the first and second phase. Each time a classification is tagged as suspicious, the agent interacts with the Interface Agent to request an expert review. Finally, this agent controls the alert mechanism and coordinates the actions required for responding to this type of attack.
- **Interface Agent:** This agent was designed to function in different devices (PDA, Laptop, Workstation). It facilitates ubiquitous communication with the security personnel when an alert has been sent by the Coordinator Agent. Additionally, it facilitates the process of adjusting the global configuration of the architecture. This agent is also located in the highest layer of the architecture.

## 4 Results and Conclusions

This article has presented a distributed hierarchical multi-agent architecture for classifying SOAP messages. The architecture was designed to exploit the distributed capacity of the agents. Additionally, an advanced classification mechanism was designed to filter incoming SOAP messages. The classification mechanism was structured in two phases, each of which includes a special CBR-BDI agent that functions as a classifier. The first phase filters simple attacks without exhausting an excessive amount of resources by applying a CBR engine that incorporates a decision tree. The second phase, a bit more complex and costly, is in charge of classifying the SOAP messages that were not classified in the first phase. This phase applies a CBR engine that incorporates a neural network. A prototype of our proposed solution was based on a classification mechanism and developed in order to evaluate its effectiveness. Figure 2 shows the results obtained for a set of SOAP messages.

Figure 2 shows the percentage of prediction with regards to the number of patterns (SOAP messages) for the classification mechanism. It is clear that as the number of patterns increases, the success rate of prediction also increases in terms of percentage. This is influenced by the fact that we are working with CBR systems, which depend on a larger amount of data stored in the memory of cases.

Future works are expected to develop the tools for obtaining a complete solution and to evaluate the successfulness in repelling and detecting attacks. With the advantage of a distributed process for classification tasks, it would be possible to evaluate the effectiveness of the classification mechanism, and the response time.

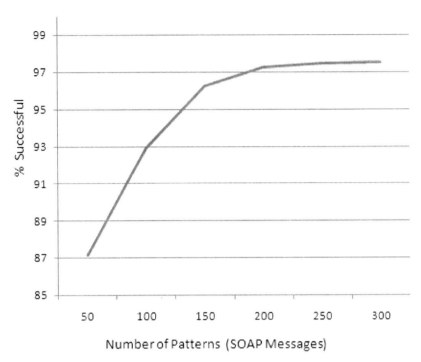

**Fig. 2.** Effectiveness of the classification mechanism integrated according to the number of patterns

**Acknowledgments.** This development has been partially supported by the Spanish Ministry of Science project TIN2006-14630-C03-03 and The Professional Excellence Program 2006-2010 IFARHU-SENACYT-Panama.

# References

1. OASIS: Web Services Security: SOAP Message Security 1.1 (WS-Security 2004), OASIS Standard 2004,
   http://docs.oasis-open.org/wss/2004/01/
   oasis-200401-wss-soap-message-security-1.0.pdf (2006)
2. Bajaj, et al.: Web Services Policy Framework, WS-Policy (2004), http://www.ibm.com/developerworks/library/specification/ws-polfram
3. Web Services Trust Language (WS-Trust),
   http://specs.xmlsoap.org/ws/2005/02/trust/WS-Trust.pdf
4. Web Services Secure Conversation Language (WS-SecureConversation), http://specs.xmlsoap.org/ws/2005/02/sc/WS-SecureConversation.pdf
5. Gruschka, N., Luttenberger, N.: Protecting Web Services from DoS Attacks by SOAP Message Validation. Security and Privacy in Dynamic Environments (201), 171–182 (2006)

6. Laza, R., Pavon, R., Corchado, J.M.: A Reasoning Model for CBR_BDI Agents Using an Adaptable Fuzzy Inference System. In: Conejo, R., Urretavizcaya, M., Pérez-de-la-Cruz, J.-L. (eds.) CAEPIA/TTIA 2003. LNCS (LNAI), vol. 3040, pp. 96–106. Springer, Heidelberg (2004)
7. Weerawarana, S., Curbera, F., Leymann, F., Storey, T., Ferguson, D.F.: Web Services Platform Architecture: SOAP. In: WSDL, WS-Policy, WS-Addressing, WS-BPEL, WS-Reliable Messaging, and More. Prentice Hall PTR, Englewood Cliffs (2005)
8. Loh, Y., Yau, W., Wong, C., Ho, W.: Design and Implementation of an XML Firewall. Computational Intelligence and Security 2, 1147–1150 (2006)
9. Yee, G., Shin, H., Rao, G.S.V.R.K.: An Adaptive Intrusion Detection and Prevention (ID/IP) Framework for Web Services. In: International Conference on Convergence Information Technology, pp. 528–534. IEEE Computer Society, Washington (2007)
10. Jensen, M., Gruschka, N., Herkenhoner, R., Luttenberger, N.: SOA and Web Services: New Technologies, New Standards - New Attacks. In: Fifth European Conference on Web Services-ECOWS 2007, pp. 35–44 (2007)
11. Dagdee, N., Thakar, U.: Intrusion Attack Pattern Analysis and Signature Extraction for Web Services Using Honeypots. In: First International Conference Emerging Trends in Engineering and Technology, pp. 1232–1237 (2008)
12. Carrascosa, C., Bajo, J., Julian, V., Corchado, J.M., Botti, V.: Hybrid multiagent architecture as a real-time problem-solving model. Expert Syst. Appl. 34, 2–17 (2008)
13. Corchado, J.M., Bajo, J., Abraham, A.: GerAmi: Improving Healthcare Delivery in Geriatric Residences. IEEE Intelligent Systems 23, 19–25 (2008)

# A Self-learning Anomaly-Based Web Application Firewall

Carmen Torrano-Gimenez, Alejandro Perez-Villegas, and Gonzalo Alvarez

Instituto de Física Aplicada, Consejo Superior de Investigaciones Científicas,
Serrano 144 - 28006, Madrid, Spain
{carmen.torrano,alejandro.perez,gonzalo}@iec.csic.es

**Abstract.** A simple and effective web application firewall is presented. This system follows the anomalous approach, therefore it can detect both known and unknown web attacks. The system decides whether the incoming requests are attacks or not aided by an XML file. The XML file contains the normal behavior of the target web application statistically characterized and is built from a set of normal requests artificially generated. Any request which deviates from the normal behavior is considered anomalous. The system has been applied to protect a real web application. An increasing number of training requests have been used to train the system. Experiments show that when the XML file has enough data to closely characterize the normal behaviour of the target web application, a very high detection rate is reached while the false alarm rate ramains very low.

## 1 Introduction

Web applications are becoming increasingly popular and complex in all sorts of environments, ranging from e-commerce applications to banking. As a consequence, web applications are subject to all sort of attacks, many of which might be devastating [1]. In order to detect web-specific attacks the detection is to be moved to the application layer.

An Intrusion Detection System (IDS) analyzes information from a computer or a network to detect malicious actions and behaviors that can compromise the security of a computer system. Traditionally, IDS's have been classified as either signature detection systems (also called negative approach) or anomaly detection systems (positive approach).

The first method looks for signatures of known attacks using pattern matching techniques against a frequently updated database of attack signatures. It is unable to detect new attacks and to work properly, databases must be updated frequently. Signature matching usually requires high computational effort.

The second method overcome these problems, however is prone to more false positives. It looks for anomalous system activity: once normal behavior is well defined, irregular behavior will be tagged as intrusive. A disadvantage is that in rather complex environments, obtaining an up-to-date and feasible picture of what "normal" network traffic should look like proves to be a hard problem.

The results of signature-based IDSs depend on the actual signature configuration for each web application, and cannot be compared with anomaly-based IDSs.

Regarding related works, [2] presents an overview of different anomaly detection techniques. Additionally, some works have been presented concerning attack detection in web traffic: [3] uses a parameter-oriented URL format and applies several anomaly-detection models to detect possible attacks. In [4] Markov chains are used for the detection. [5] is an anomaly-based system which infers the type of the request parameters and combines different techniques to detect attacks.

In this paper, a simple and effective anomaly-based Web Application Firewall (WAF) is presented. This system relies on an XML file to describe what a normal web application is. Any irregular behavior is flagged as intrusive. The XML file must be tailored for every target application to be protected.

The rest of the paper is organized as follows. In Sect. 2, a system overview is given, where system architecture, normal behavior modeling, and attack detection are explained. Section 3 refers to experiments. Traffic generation, the training phase, the test phase and results are also described. Section 4 describes system limitations and suggests future work. Finally, in Sec. 5, the conclusions of this work are captured.

## 2 System Overview

### 2.1 Architecture

Our anomaly-based detection approach analyzes HTTP requests sent by a client browser trying to get access to a web server. The analysis takes place exclusively at the application layer. Thus, the system can be seen as an anomaly-based Web Application Firewall (WAF), in contrast with existing signature-based WAFs [6].

In our architecture, the system operates as a proxy located between the client and the web server. Likewise, the system might be embedded as a module within the server. However, the first approach enjoys the advantage of being independent of the web platform. A diagram of the system's architecture is shown in Fig. 1.

The input of the system consists of a collection of HTTP requests $\{r_1, r_2, \ldots r_n\}$. The output is a single bit $a_i$ for each input request $r_i$, which indicates whether the request is normal or anomalous. The proxy is able to work in two different modes of operation: as IDS and as firewall.

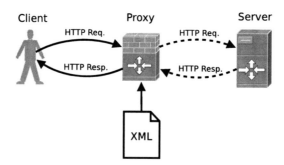

**Fig. 1.** Web Application Firewall Architecture

In detection mode, the proxy simply analyzes the incoming packets and tries to find suspicious patterns. If a suspicious request is detected, the proxy launches an alert; otherwise, it remains inactive. In any case, the request will reach the web server. When operating in detection mode, attacks could succeed, whereas false positives don't limit the system functionality.

In firewall mode, the proxy receives requests from clients and analyzes them. If the request is valid, the proxy routes it to the server, and sends back the received response to the client. If not, the proxy blocks the request, and sends back a generic denegation access page to the client. Thus, the communication between proxy and server is established only when the request is deemed as valid.

## 2.2 Normal Behavior Description

Prior to the detection process, the system needs a precise picture of what the normal behavior is in a specific web application. For this purpose, our system relies on an XML file which contains a thorough description of the web application's normal behavior. Once a request is received, the system compares it with the normal behavior model. If the difference exceeds the given thresholds, then the request is flagged as an attack and an alert is launched.

The XML file contains rules regarding to the correctness of HTTP verbs, HTTP headers, accessed resources (files), arguments, and values for the arguments. This file contains three main nodes:

**Verbs.** The *verbs* node simply specifies the list of allowed HTTP verbs. Requests using any other verb will be rejected.

**Headers.** The *headers* node specifies a list of some HTTP headers and their allowed values. Different values will not be accepted.

**Directories.** The *directories* node has a tree-like structure, in close correspondence to the web application's directory structure.

1. Each directory in the web application space is represented in the XML file by a *directory* node, allowing nesting of directories within directories. The attribute *name* defines these nodes.
2. Each file in the web application space is represented by a *file* node within a *directory* node and is defined by its attribute *name*.
3. Input arguments are represented by *argument* nodes within the corresponding *file* node. Each argument is defined by its name and a boolean value *requiredField* indicating whether the request should be rejected if the argument is missing.
4. Legal values for arguments should meet some statistical rules, which are represented by a *stats* node within the corresponding *argument* node. These statistical properties together give a description of the expected values.

    Each relevant property is defined by an attribute within the *stats* node. In our approach we considered the following relevant properties:
    - *Special*: set of special characters (no letters and no digits) allowed
    - *lengthMin*: minimum input length

- *lengthMax*: maximum input length
- *letterMin*: minimum number of letters
- *letterMax*: maximum number of letters
- *digitMin*: minimum number of digits
- *digitMax*: maximum number of digits
- *specialMin*: minimum number of special characters
- *specialMax*: maximum number of special characters

These properties allow to define four intervals (length, letters, digits and special characters) of the allowed values for each argument. Requests with argument values exceeding their corresponding normal intervals will be rejected.

The adequate construction of the XML file with the suitable intervals is crucial for a good detection process. An example of XML configuration file is shown in Fig. 2.

```
<configuration>
<verbs>
  <verb>GET</verb>
  <verb>POST</verb>
</verbs>
<headers>
  <rule name="Accept-Charset"
        value="ISO-8859-1"/>
</headers>
<directories>
  <directory name="shop">
    <file name="index.jsp"/>
    <directory name="public">
      <file name="add.jsp">
        <argument name="quantity" requiredField="true">
          <stats maxDigit="100" maxLength="1" maxLetter="0"
                 maxOther="0" minDigit="100" minLength="1"
                 minLetter="0" minOther="0" special=""/>
        </argument>
        ...
```

**Fig. 2.** XML file example

### 2.3 Detection Process

The detection process is a sequence of tests where the different parts of the incoming requests are checked with the aid of the generated XML file. If an incoming request fails to pass one of these tests, an attack is assumed. Following are the steps of the detection process:

1. Verb check. The verb must be present in the XML file.
2. Headers check. If the header appears in the XML file, its value must be included too.
3. Resource test. Only files present in the XML file are allowed.

4. Arguments test (if any):
   (a) It is checked that all arguments are allowed for the resource
   (b) It is confirmed that all mandatory arguments are present in the request
   (c) Argument values are cheched. If any propertie of the argument is out of the corresponding interval or contains any forbidden special character, the request is considered anomalous.

# 3 Experiments

## 3.1 Case Study: Web Shopping

The WAF has been configured to protect a specific web application, consisting of an e-commerce web store, where users can register and buy products using a shopping cart.

## 3.2 XML File Generation

As already stated, the XML file describes the normal behavior of the web application. Therefore, to train the system and configure this file, only normal and non-malicious traffic to the target web application is required. Nevertheless, how to obtain only normal traffic may not be an easy task. To obtain good detection results thousands of requests are needed. There are some alternatives to obtain normal traffic:

1. Thousands of legitimate users can surf the target web application and generate normal traffic. However, getting thousands of people to surf the web might not be easy.
2. The application can be published in the Internet. Unfortunately, attacks will be mixed with normal traffic, so this traffic cannot be used to train the system. This is the approach followed in [3].
3. Traffic can be generated artificially. Although the traffic is not real, we can be sure that only normal traffic is included. Hence we considered this alternative the most suitable for our purposes.

## 3.3 Artificial Traffic Generation

In our approach, normal and anomalous request databases were generated artificially with the help of dictionaries.

**Dictionaries.** Dictionaries are data text which contain real data to fill the different arguments used in the target application. All the dictionaries used (names, surnames, addresses, etc.) were extracted from real databases.

A set of dictionaries containing only allowed values was used to generate the normal request database, and a different set containing attacks and illegal values was used to generate the anomalous request database.

**Normal Traffic Generation.** Allowed HTTP requests were generated for each page in the web application. Arguments and cookies in the page, if any, were also filled out with values from the normal dictionaries. The result was a normal request database (*NormalDB*), which was used both in the training and test phase.

**Anomalous Traffic Generation.** Illegal HTTP requests were generated with the help of anomalous dictionaries. The result was an anomalous request database (*AnomalousDB*), which was used only in the test phase. Three types of anomalous requests were considered:

1. **Static attacks** fabricate the resource requested. These requests include obsolete files, session id in URL rewrite, directory browsing, configuration files, and default files.
2. **Dynamic attacks** modify valid request arguments: SQL injection, CRLF injection, cross-site scripting, server side includes, buffer overflows, etc.
3. **Unintentional illegal requests**. These requests should also be rejected even though they do not have malicious intention.

### 3.4 Training Phase

During the training phase, the system learns the web application normal behavior. The aim is to obtain the XML file from normal requests. In the construction of the XML file, different HTTP aspects must be taken into account.

– Argument values are characterized by extracting their statistical properties from the requests.
– Verbs, resources and certain headers found in the requests are included directly in the XML file as allowed elements.

### 3.5 Test Phase

During the test phase the proxy accepts requests from both databases, NormalDB and AnomalousDB, and relies on the XML file to decide whether the requests are normal or anomalous.

The performance of the detector is then measured by Receiver Operating Characteristic (ROC) curves [7]. A ROC curve plots the attack detection rate (true positives, $TP$) against the false alarm rate (false positives, $FP$).

$$DetectionRate = \frac{TP}{TP+FN} \qquad (1)$$

$$FalseAlarmRate = \frac{FP}{FP+TN} \qquad (2)$$

In order to obtain diffetent points in the ROC curve, the number of training requests was used as a tuning parameter.

### 3.6 Results

Several experiments have been performed using an increased amount of training requests in the training phase. For each experiment, the proxy received 1000 normal requests and 1000 attacks during the test phase.

Figure 3 shows the results obtained by the WAF while protecting the tested web application. As can be seen, very satisfactory results are obtained: the false alarm rate is close to 0 whereas the detection rate is close to 1. At the beginning, with a low amount of training requests, the proxy rejects almost all requests (both normal and attacks). As a consequence, the detection rate is perfect (1) and the false positive rate is high. As the training progresses, the false alarm rate decreases quickly and the detection rate remains reasonably high.

It is important to notice that when the XML file closely characterizes the web application normal behaviour, the different kinds of attacks can be detected and few false alarms are raised.

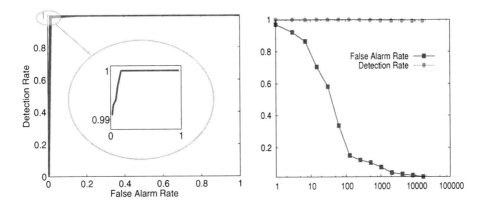

**Fig. 3.** On the left, the ROC curve of WAF protecting the web store is presented. On the right, the false alarm rate and the detection rate vs the number of training requests is plotted.

## 4 Limitations and Future Work

As shown in the previous section, when the XML file is configured correctly, the system succeeds in detecting any kind of web attacks. Thus, the main issue is how to automatically configure the XML description file. In our approach, the XML file is built from a set of allowed requests in the target web application. However, obtaining only normal traffic may not be an easy task, as was discussed in Sec. 3.2. Therefore, the main limitation consists in correctly implementing the training phase for any web application.

Other limitations arise when protecting complex web applications. For instance, web sites that create and remove pages dynamically, generate new URLs to access resources, or allow users for updating contents, may difficult the XML file configuration. Further modifications of the system will attempt to solve these problems.

Future work refers to signing cookies and hidden fields in order to avoid cookie poisoning and hidden field manipulation attacks. Also, URL patterns will be used in describing sites with dynamic resources.

## 5 Conclusions

We presented a simple and efficient web attack detection system or Web Application Firewall (WAF). As the system is based on the anomaly-based methodology it proved to be able to protect web applications from both known and unknown attacks. The system analyzes input requests and decides whether they are anomalous or not. For the decision, the WAF relies on an XML file which specifies web application normal behavior. The experiments show that as long as the XML file correctly defines normality for a given target application, near perfect results are obtained. Thus, the main challenge is how to create an accurate XML file in a fully automated manner for any web application. We show that inasmuch great amounts of normal (non-malicious) traffic are available for the target application, this automatic configuration is possible using a statistical characterization of the input traffic.

## Acknowledgements

We would like to thank the Ministerio de Industria, Turismo y Comercio, project SE-GUR@ (CENIT2007-2010), project HESPERIA (CENIT2006-2009), the Ministerio de Ciencia e Innovacion, project CUCO (MTM2008-02194), and the Spanish National Research Council (CSIC), programme JAE/I3P.

## References

1. Alvarez, G., Petrovic, S.: A new taxonomy of Web attacks suitable for efficient encoding. Computers and Security 22(5), 453–449 (2003)
2. Patcha, A., Park, J.: An overview of anomaly detection techniques: Existing solutions and latest technological trends. Computer Networks 51(12), 3448–3470 (2007)
3. Kruegel, C., Vigna, G., Robertson, W.: A multi-model approach to the detection of web-based attacks. Computer Networks 48(5), 717–738 (2005)
4. Estévez-Tapiador, J., García-Teodoro, P., Díaz-Verdejo, J.: Measuring normality in HTTP traffic for anomaly-based intrusion detection. Computer Networks 45(2), 175–193 (2004)
5. Bolzoni, D., Zambon, E.: Sphinx: An anomaly-based web intrusion detection system. In: Workshop on Intrusion Detection Systems, Utrecht, The Netherlands, 14 pages (2007)
6. ModSecurity. Open Source signature-based Web Application Firewall (2009), http://www.modsecurity.org
7. Provost, F., Fawcett, T., Kohavi, R.: The case against accuracy estimation for comparing induction algorithms. In: Proceedings of the 15th International Conference on Machine Learning. Morgan Kaufmann, San Francisco (1998)

# An Investigation of Multi-objective Genetic Algorithms for Encrypted Traffic Identification

Carlos Bacquet, A. Nur Zincir-Heywood, and Malcolm I. Heywood

Dalhousie University, Faculty of Computer Science
{bacquet,zincir,mheywood}@cs.dal.ca

**Abstract.** The increasing use of encrypted traffic combined with non-standard port associations makes the task of traffic identification increasingly difficult. This work adopts a multi-objective clustering approach to the problem in which a Genetic Algorithm performs both feature selection and cluster count optimization under a flow based representation. Solutions do not use port numbers, IP address or payload. Performance of the resulting model provides 90% detection 0.8% false positive rates with 13 clusters supported by 14 of the original 38 features.

## 1 Introduction

An important part of network management requires the accurate identification and classification of network traffic [3]. Network administrators are normally interested in identifying application types for decisions regarding both bandwidth management and quality of service [3]. A particularly interesting area in network traffic identification pertains to encrypted traffic, where the fact that the payload is encrypted represents an additional degree of uncertainty. Specifically, many traditional approaches to traffic classification rely on payload inspection, which become unfeasible under packet encryption [8, 10, 5, 4, 9]. An alternative to payload inspection would be the use of port numbers to identify application types. However, this practice has become increasingly inaccurate, as users are now able to arbitrarily change the port number to deceive security mechanisms [3, 10, 5, 9, 4]. In short, the traditional approaches are unable to deal with the identification of encrypted traffic.

In this work our objective is the identification of encrypted traffic, where Secure Shell (SSH) is chosen as an example encrypted application. While SSH is typically used to remotely access a computer, it can also be utilized for "tunneling, file transfers and forwarding arbitrary TCP ports over a secure channel between a local and a remote computer" [3]. These properties of SSH make it an interesting encrypted application to focus on. From the traffic identification perspective a Multi-Objective Genetic Algorithm (MOGA) is pursued; where this facilitates the dual identification of appropriate (flow) feature subspace and clustering of traffic types. Such a system assumes that the resulting clusters partition traffic into encrypted/not encrypted. Performance of the system achieves detection rates above 90% and false positive rates below 1%.

## 2 Previous Work

Given the limitations of port number analysis and payload inspection, several previous attempts to identify encrypted traffic have worked with the statistics of the flow [3, 10, 5]. A number of these attempts have employed supervised learning methods. However, these classifiers have uncertain generalization properties when faced with new data [10, 5]. One alternative to classifiers is the use of clustering mechanisms or unsupervised learning methods. The following is an analysis of previous work with unsupervised methods. To the best of our knowledge, this is the first work that utilizes a genetic algorithm for the dual problem of feature selection and clustering for encrypted traffic identification.

Recently, Siqueira *et al.* presented a clustering approach to identify Peer-to-Peer (P2P) versus non-P2P in TCP traffic [8]. A total of 249 features from the headers of the packets were considered, out of which the best 5 were selected to parameterize clusters. The selection of the features was based on the variance of their values, such that the higher the variance, the better the perceived discrimination of a feature. This methodology enabled the authors to achieve results of 86.12% detection rate for P2P applications with an average accuracy of 96.79%. However, the method utilized the port number as one of the five features. Erman *et al.* also presented a clustering approach for the network traffic identification problem [5]. They evaluated three clustering algorithms: K-means, DBSCAN and AutoClass. The authors used two datasets, one of them (Auckland IV) containing DNS, FTP, HTTP, IRC, LimeWire, NNTP, POP3, and SOCKS; and the other (Calgary trace), containing HTTP, P2P, SMTP, and POP3. The authors found that with K-means the overall accuracy steadily improved as the number of clusters was increased. This continued until $K$ was around 100 with the overall accuracy being 79% and 84% on each data set respectively. Thus, from their results they observed that K-means seemed to provide the best mix of properties. Bernaille *et al.* used information from the first five packets of a TCP connection to identify applications [4]. In particular, they analyzed the size of the first few packets, which captures the application's negotiation phase. They then used a K-means algorithm to cluster these features during the learning phase. After several experiments they concluded that the best $K$ number of clusters was 50, achieving results of up to 96.9% accuracy with SSH. The authors did mention that the method is sensitive to packet order, thus a potential limitation. Yingqiu *et al.* also presented a flow based clustering approach, using K-means to build the clusters with features previously identified as the best discriminators [10]. To this end, several feature selection and search techniques are performed. After clustering, the authors applied a log transformation, which enabled them to reach an overall accuracy level of up to 90% when utilizing $K = 80$ clusters. The authors concluded this was a very promising approach, however, they also mention that they only worked with TCP. It should be noted here that none of the above work reported their false positive rates.

## 3 Methodology

In this section we first characterize the data utilized for our training and testing. The entire process employed in this work is outlined in Figs. 1 and 2. The data set used was captured by the Dalhousie University Computing and Information Services Centre (UCIS) in January 2007 on the campus network between the university and the commercial Internet. Given the privacy related issues the university may face, the data was filtered to scramble the IP addresses and each packet was further truncated to the end of the IP header so that all the payload was excluded. Furthermore, the checksums were set to zero since they could conceivably leak information from short packets. However, any length information in the packet was left intact. Dalhousie traces were labeled by a commercial classification tool called PacketShaper, i.e., a deep packet analyzer [14]. PacketShaper uses Layer 7 filters (L7) to classify the applications. Given that the handshake part of SSH protocol is not encrypted, we can confidently assume that the labeling of the data set is 100% correct and provides the ground truth for our data set. We emphasize that our work did not consider any information from the handshake phase nor any part of the payload, IP addresses or port numbers. The handshake phase was used solely to obtain the ground truth to which we compare our obtained results. Also, we focus on SSH as a case study, we could have employed any other encrypted traffic protocol. However, the fact the SSH's handshake is not encrypted, allowed us to compare our obtained results with those obtained through payload inspection. In order to build training data and test data, we divided the trace into two partitions. We sampled our training data from the first partition and the test data from the remaining partition. The training data consisted of 12240 flows, including SSH, MSN, HTTP, FTP, and DNS. The test data, on the other hand, consisted of 1000 flows of each of those applications, plus 1000 flows that belonged to any of the following applications: RMCP, Oracle SQL*NET, NPP, POP3, NETBIOS Name Service, IMAP, SNMP, LDAP, NCP, RTSP, IMAPS and POP3S.

Flows are defined by sequences of packets that present the same values for source IP address, destination IP address, source port, destination port and type of protocol [8]. Each flow is described by a set of statistical features and associated feature values. A feature is a descriptive statistic that can be calculated from one or more packets. We used the NetMate [11] tool set to process data sets, generate flows, and compute feature values. In total, 38 features were obtained. Flows are bidirectional with the first packet determining the forward direction. Since flows are of limited duration, in this work UDP flows are terminated by a flow timeout, and TCP flows are terminated upon proper connection teardown or by a flow timeout, whichever occurs first. Given that there is not a general and accepted value for the flow time out, a 600 second flow timeout value was employed here; where this corresponds to the IETF Realtime Traffic Flow Measurement working groups architecture [13]. It is important to mention that only UDP and TCP flows are considered. Specifically, flows that have no less than one packet in each direction, and transport no less than one byte of payload.

As for feature selection we have employed the model proposed by YeongSeog et al. [6] and modified its evolutionary component to follow the model proposed by Kumar et al. [7]. The latter ensures a convergence towards the Pareto-front (set of non-dominated solutions) without any complex sharing/niching mechanism. One specific property of this Genetic Algorithm (GA) is the utility of a steady-state GA, thus only one or two members of the population are replaced at a time, Fig. 2. A GA starts with a population of individuals (potential solutions to a problem), and incrementally evolves that population into better individuals, as established by the fitness criteria. Fitness is naturally relative to the population. Then, for several iterations, individuals are selected to be combined (crossover) to create new individuals (offspring) under a fitness proportional selection operator. In order to model the problem of feature selection to the GA, each individual in the population represents a subset of features $f$ and a number of clusters $K$. Specifically, an individual is a 60 bit binary string, with the first 38 bits representing the features to include and the remaining 22 bits representing the $K$ number of clusters. Bits of the individuals in the initial population are initialized with a uniform probability distribution. For the feature selection, a one implies to include the feature at that index, and a zero means to discard it. The K number of clusters, on the other hand, is represented by the number of "ones" (as opposed to "zeros") contained between the $39^{th}$ bit and the $60^{th}$ bit, plus two. The reason for adding two is that zero or one cluster would never be a solution. Clusters are identified using the standard K-means algorithm, using that subset of features $f$, and the number of clusters $K$, as the input for the K-means algorithm. For this step we used the K-means algorithm provided by Weka [12]. The fitness of the individual will then depend on how well the resulting clusters perform in relation to the following four predefined clustering objectives:

- *Fwithin* (measures cluster cohesiveness, the more cohesive the better). We calculate the average standard deviation per cluster. That is, per each $i$'th cluster, the sum of the standard deviations per feature over the total number of employed features. Then *Fwithin* will be $K$ over the sum of all the clusters' average standards.

- *Fbetween* (measures how separate the clusters are from each other, the more separated the better). For each pair of cluster $i$ and $j$ we calculate its average standard deviations and we also calculate the euclidean distance between its centroids. Then, *Fbetween* for clustetr $i$ and $j$ is:

$$Fbetween\_i\text{-}j = \frac{Euclidean\ distance\ from\ i\ to\ j}{\sqrt{(AveStdDev\_i)^2 + (AveStdDev\_j)^2}} \quad (1)$$

Thus, Fbetween will be the sum of all pairs of cluster's *Fbetween_i-j*, over K.

- *Fclusters* (measures the number of clusters $K$, "Other things being equal, fewer clusters make the model more understandable and avoid possible over fitting"[6])

$$Fclusters = 1 - \frac{K - Kmin}{Kmax - Kmin} \quad (2)$$

*Kmax* and *Kmin* are the maximum and minimum number of clusters.

- *Fcomplexity* (measures the amount of features used to cluster the data, this objective aims at minimizing the number of selected features)

$$Fcomplexity = 1 - \frac{d-1}{D-1} \quad (3)$$

$D$ is the dimensionality of the whole dataset and $d$ is the number of employed features.

Instead of combining these objectives into a single objective, this model followed a multi-objective approach, which has the goal to approximate to the Pareto front, or set of non-dominated solutions. Informally, a solution is said to dominate another if it has higher values in at least one of the objective functions (*Fwithin*, *Fbetween*, *Fclusters*, *Fcomplexity*), and is at least as good in all the others. After the objective values for each individual have been assessed, individuals are assigned with ranks, which indicate how many individuals dominate that particular individual. Finally, the fitness of the individuals is inversely proportional to their ranks, which is used to build a roulette wheel that is ultimately used for parental selection under the aforementioned steady state model. The initial population is evolved for 5000 epochs, after which we consider the set of non-dominated individuals from it (individuals whose ranks equal to 1). These individuals correspond to the set of potential solutions. The evolutionary component of the algorithm is then terminated and the best individual in the set of non-dominated solutions (the one that better identifies SSH traffic) is identified. We take each individual from the set of non dominate solutions and label its clusters as either SSH or non-SSH, based on the labels in the training data. If the majority of the flows in a cluster have SSH labels, then that cluster is labeled as SSH, otherwise it is labeled as non-SSH. Then, the post-training phase consists of testing each individual in our labeled training data (used to build the clusters on), to identify the solution with best classification rate, which will be the final solution.

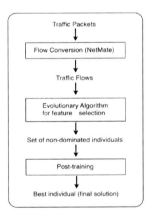

**Fig. 1.** System Diagram

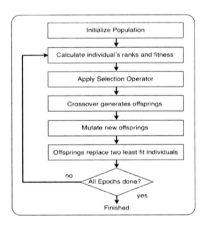

**Fig. 2.** Evolutionary Component Diagram

## 4 Results

In traffic classification, two metrics are typically used to quantify performance: Detection Rate (DR), Eq. 4, and False Positive Rate (FPR), Eq 5. In this case, detection rate will reflect the number of SSH flows correctly classified, and FPR will reflect the number of Non-SSH flows incorrectly classified as SSH. As we can observe, a high DR and a low FPR would be the desired outcomes.

$$DR = 1 - \frac{\#\ false\ negatives}{total\_number\_of\_SSH\_flows} \qquad (4)$$

$$FPR = \frac{\#\ false\ positives}{total\_number\_of\_non\_SSH\_flows} \qquad (5)$$

*false_negatives* mean SSH traffic incorrectly classified as non-SSH traffic. In these experiments we compared our proposed system, MOGA, with K-means algorithm using the same feature set and the same data sets. Weka's K-means clusters were built with values of $K$ from 5 to 100. The results show more than 94% DR with very high FPR (more than 3.4%). Table 1.

**Table 1.** Baseline – K-means on Dalhousie data sets

| K | DR | FPR | K | DR | FPR | K | DR | FPR | K | DR | FPR |
|---|---|---|---|---|---|---|---|---|---|---|---|
| 5 | 0.94 | 0.12 | 25 | 0.95 | 0.04 | 45 | 0.97 | 0.10 | 80 | 0.98 | 0.12 |
| 10 | 0.97 | 0.16 | 30 | 0.97 | 0.16 | 50 | 0.97 | 0.15 | 90 | 0.97 | 0.06 |
| 15 | 0.97 | 0.18 | 35 | 0.97 | 0.16 | 60 | 0.97 | 0.07 | 100 | 0.97 | 0.07 |
| 20 | 0.95 | 0.04 | 40 | 0.96 | 0.034 | 70 | 0.98 | 0.14 | | | |

On the other hand, the MOGA was run 25 times, producing a combined of 783 non-dominated individuals. Out of those individuals, we identified in the post-training phase the one that had the lowest FPR and a DR close to 90%. We considered that individual to be our final solution. Fig. 3 displays the plot of the candidate solutions in the post-training phase. The *x*-axis represents the FPR and the *y*-axis represents the DR. The selected individual, which is represented by a larger black square instead of a gray diamond, achieved a detection rate of 90% and a false positive rate of 0.08% in the post training phase. That same individual achieved a detection rate of 92% and a false positive rate of 0.8% in the test data. This final solution employed only 14 out of the 38 available features, and it clustered the data into 13 clusters. Thus, this solution not only achieved very promising results in terms of detection rate and false positives rate, but it also considerably decreased the number of employed features, and clustered the data with a relatively low value of $K$ clusters, both very desirable outcomes. The features employed by our final solution are displayed in Table 2. In addition, an analysis of the best 10 individuals reveals that these other well performing individuals also employed some of these features. Table 3 displays which features were the most recurrent among the top ten individuals.

Finally, it is also important to mention the frequency with which we were able to obtain these kinds of results in our conducted runs. Given the stochastic component of evolutionary algorithms, it is expected that some runs will be better than others. In this case, we were able to obtain individuals with detection rates above 90% and false positive rates below 1% in 21 out of the 25 conducted runs.

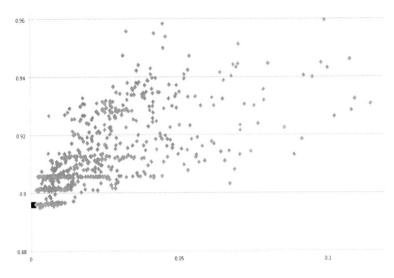

**Fig. 3.** Plot of Detection rate versus false positives in post-training phase

**Table 2.** Features employed by the best individual

| Feature | Feature name |
|---|---|
| Total back packets (total_bpackets) | Min back inter arrival time (min_biat) |
| Total back volume (total_bvolume) | Mean back inter arrival time (mean_biat) |
| Mean fwrd packet length (mean_fpktl) | Max back inter arrival time (max_biat) |
| Max fwrd packet length (max_fpktl) | Sub-flow fwrd packets (sflow_fpackets) |
| Std fwrd packet length (std_fpktl) | Sub-flow fwrd bytes (sflow_fbytes) |
| Min back packet length (min_bpktl) | Back push counters (bpsh_cnt) |
| Std fwrd inter arrival time (std_fiat) | Fwrd urgent counters (furg_cnt) |

**Table 3.** Most recurrent features employed by top ten individuals

| Feature | # | Feature | # |
|---|---|---|---|
| Min back inter arrival time (min_biat) | 10 | Max back packet length (max_bpktl) | 5 |
| Total fwrd packets (total_fpackets) | 7 | Min fwrd inter arrival time (min_fiat) | 5 |
| Total back packets (total_bpackets) | 7 | Std fwrd inter arrival time (std_fiat) | 5 |
| Std back inter arrival time (std_biat) | 7 | Total back volume (total_bvolume) | 5 |
| Fwrd urgent counters (furg_cnt) | 7 | Sub-flow fwrd packets (sflow_fpackets) | 5 |
| Mean fwrd packet length (mean_fpktl) | 6 | Sub-flow back packets (sflow_bpackets) | 5 |
| Std fwrd packet length (std_fpktl) | 5 | Back urgent counters (burg_cnt) | 5 |

# 5 Conclusions

A steady state MOGA is applied to the dual problem of feature subspace selection and clustering of encrypted traffic. Compared to basic K-means algorithm, the proposed method not only considerably decreases the number of features needed for clustering the flows, but it also made it possible to cluster the data with a very small value of $K$.

Both of which are very desirable advantages for a potential implementation of an encrypted traffic detection system. The results were presented in terms of DR and FPR. Our best individual achieved a detection rate of 92% and a false positive rate of 0.8%, whereas it employed only 14 out of the 38 available features and clustered the data in only 13 clusters. For future work, we aim to apply this methodology to other types of encrypted traffic such as Skype.

## References

[1] Alshammari, R., Zincir-Heywood, N.: Generalization of Signatures for SSH Traffic Identification. In: IEEE Symposium Series on Computational Intelligence (2009)
[2] Alshammari, R., Zincir-Heywood, N.: Investigating two different approaches for encrypted traffic classification. In: Sixth Annual Conference on Privacy, Security and Trust, pp. 156–166 (2008)
[3] Alshammari, R., Zincir-Heywood, N.: A flow based approach for SSH traffic detection. ISIC. In: IEEE International Conference on Systems, Man and Cybernetics, pp. 296–301 (2007)
[4] Bernaille, L., Teixeira, R., Akodkenou, I., Soule, A., Salamatian, K.: Traffic classification on the fly. SIGCOMM Comput. Commun. Rev. 36(2), 23–26 (2006)
[5] Erman, J., Arlitt, M., Mahanti, A.: Traffic classification using clustering algorithms. In: SIGCOMM Workshop on Mining Network Data, pp. 281–286. ACM, New York (2006)
[6] YeongSeog, K., Street, W.N., Menczer, F.: Feature selection in unsupervised learning via evolutionary search. In: Sixth ACM SIGKDD International Conference on Knowledge Discovery and Data Mining, pp. 365–369. ACM, New York (2000)
[7] Kumar, R., Rockett, P.: Improved sampling of the pareto-front in multiobjective genetic optimizations by steady-state evolution: A pareto converging genetic algorithm. Evol. Comput. 10(3), 283–314 (2002)
[8] Siqueira Junior, G.P., Bessa Maia, J.E., Holanda, R., Neuman de Sousa, J.: P2P traffic identification using cluster analysis. In: First International Global Information Infrastructure Symposium (GIIS), pp. 128–133 (2007)
[9] Wright, C., Monrose, F., Masson, G.: On inferring application protocol behaviors in encrypted network traffic. J. Mach. Learn. Res. 7, 2745–2769 (2006)
[10] Yingqiu, L., Wei, L., Yun-Chun, L.: Network traffic classification using K-means clustering. In: Second International Multi-Symposiums on Computer and Computational Sciences (IMSCCS), pp. 360–365. IEEE Computer Society, Washington (2007)
[11] NetMate, http://www.ip-measurement.org/tools/netmate/
[12] WEKA Software, http://www.cs.waikato.ac.nz/ml/weka/
[13] IETF,
http://www3.ietf.org/proceedings/97apr/
97apr-final/xrtftr70.htm
[14] PacketShaper, http://www.packeteer.com/products/packetshaper

# A Multi-objective Optimisation Approach to IDS Sensor Placement

Hao Chen[1], John A. Clark[1], Juan E. Tapiador[1], Siraj A. Shaikh[2], Howard Chivers[2], and Philip Nobles[2]

[1] Department of Computer Science
University of York
York YO10 5DD, UK
[2] Department of Informatics and Sensors
Cranfield University
Shrivenham SN6 8LA, UK

**Abstract.** This paper investigates how intrusion detection system (IDS) sensors should best be placed on a network when there are several competing evaluation criteria. This is a computationally difficult problem and we show how Multi-Objective Genetic Algorithms provide an excellent means of searching for optimal placements.

## 1 Introduction

Effective intrusion detection for almost any large network will require multiple sensors. However, determining where to place a set of sensors to create cost effective intrusion detection is a difficult task. There may be several evaluation criteria for placements, seeking to maximise various desirable properties (e.g. various attack detection rates), whilst seeking to reduce undesirable properties (such as false alarm rates as well as purchase, management, and communications costs). Subtle tradeoffs may need to be made between the properties; different placements may have complementary strengths and weaknesses, with neither placement being uniformly better than the other. However, engineering regularly deals with such difficult *multi-criteria optimisation* problems and has developed a powerful suite of technical tools to facilitate the search for high performing solutions. In this paper we show how a multi-objective genetic algorithm (MOGA) can be harnessed to address the sensor placement problem.

The optimal placement of sensors depends on what we wish to achieve. A placement may be optimal for the detection of one type of attack, but not for a second type of attack. We may seek a placement that gives good chances of detecting each of several types of attack; this may yield a different optimal placement. To determine the "optimal" placement we need a means to evaluate a particular placement. In some cases, this may be carried out with respect to statically assigned information (e.g. location of firewalls and servers). In others, we may need to simulate attacks and measure the effectiveness of the placement. Thus the specific evaluation mechanism may differ but the overall technique remains the same: find a placement $P$ that optimises some evaluation function $f(P)$, or a set of evaluation functions $f_1(P), \ldots, f_n(P)$. Such a situation is a suitable target for the application of heuristic optimisation.

The Genetic Algorithm (GA) [1] is a heuristic optimisation technique based loosely on natural selection and has been applied successfully in the past to a diverse set of problems. Its general idea is that populations evolve according to rules that will in general support the emergence of ever fitter individuals (that is, ones with higher evaluation value). As with other search methods, GA can be used in conjunction with Multi-Objective Optimisation (MOO) techniques [2]. MOO aims to find solutions that satisfy more than one objective, so that a solution's ability to solve a problem is assessed by a set of objective functions $f_1, \ldots, f_n$. MOO methods return a set of solutions in a single run, and each solution achieves a different balance between multiple objectives. In this paper, we experiment with GA and MOO to evolve optimal sensor placements. These experiments serve as proof of concept and to demonstrate the validity and potential of the proposed approach. Researchers have used Genetic Programming (GP) and Grammatical Evolution to determine IDS detection rules [3], but our experiments reported here report the first use of heuristic optimisation techniques to evolve optimal IDS sensor placements.

## 2 Related Work

Noel and Jajodia [4] propose to use attack graph analysis to find out optimal placement of IDS sensors. Attack graphs represent a series of possible paths taken by potential intruders to attack a given asset. Such graphs are constructed in a topological fashion taking into account both vulnerable services that allow nodes to be exploited and used as launch pads, and protective measures deployed to restrict connectivity. The purpose is to enumerate all paths leading to given assets and where optimal placement is devised to monitor all paths using minimal number of sensors. This is seen as a set cover problem: each node allows for monitoring of certain graph edges and the challenge is to find a minimum set of routers that cover all edges in the graph; a greedy algorithm is then used to compute optimal placement. The use of attack graphs provides an efficient mapping of network vulnerabilities in the network. A vulnerability-driven approach to deploying sensors overlooks factors such as traffic load however. As a result the placement is optimised such that the more paths that go through a node the more likely it is chosen for placement. Rolando [5] introduces a formal logic-based approach to describe networks, and automatically analyse them to generate signatures for attack traffic and determine placement of sensors to detect such signatures. Their notation to model networks is simple yet expressive to specify network nodes and interconnecting links in relevant detail. While there are advantages to using a formal model, such an approach may not be scalable. The formal notation allows for a more coarse-grained specification but it is not clear whether the resulting sensor configurations are even likely to be feasible for real environments. Moreover, the notation does not allow for modelling any system-level characteristics.

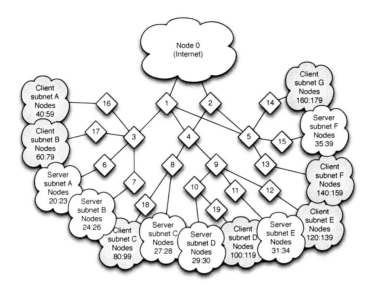

**Fig. 1.** Simulated Network

## 3 Experimental Setup and Evaluation

### 3.1 Network Simulation

We use Network Simulator NS2 [6] to simulate our experimental network as shown in Figure 1. The whole network consists of 180 nodes, where node 0 represents the outside world, nodes 1 to 19 are the routers interconnecting various parts of the network, nodes 20 to 39 are servers offering valuable services to users and therefore critical assets that need to be protected, and nodes 40 to 180 are ordinary clients some of which may be compromised by intruders to attack critical assets. The network is organised as such that the servers are distributed over six subnets and the clients are distributed over seven separate subnets.

We simulate real intrusive behaviour to analyse how such behaviours could be efficiently detected by the proposed approach. The intrusive behaviour we simulated is to do with probing and information gathering, the purpose of which is to assess a potential target's weaknesses and vulnerabilities [7].

For example, an intruder may strive to detect active hosts and networks that are reachable and the services they are running that could be successfully exploited. Detecting and preventing such probes therefore is important both to inhibit exposure of information and prevent attacks that follow.

We simulate a probe attack scenario where various servers are probed from the outside (through node 0) and inside from clients, hence the simulation consists of both external and internal attacks. An intruder may subvert a node in any of the client subnets to probe any of the servers. Intruders (picked randomly) from each of the client subnets, that are client nodes 45, 78, 95, 111, 133, 157 and 178, probe server nodes 20 to 38. In addition, node 45 also attempts a probe on neighbours 46 and 47. A total number of of 154 instances of probe attack are injected.

Note the simulation of attacks so far in our experiments is simple for the purposes of demonstration. Simulation of more subtle intrusive behaviours and research of how such behaviours could be effectively and efficiently detected by our approach are currently under investigation.

In order to investigate how the false alarms may influence sensor placement strategy, we simulate not only a number of attacks but also background network traffic. The background network traffic is generated randomly by NS2 traffic source generator *cbrgen*. In the experiment, we assume that traditional IDS metrics such as false positive rate and false negative rate are already known. This hypothesis stands as all IDS evaluation work so far is trace-driven [8], suggesting when evaluating IDSs, we use a data set where we know the ground truth, i.e., what data are attacks and what data are normal. Thus we can easily find out the metrics such as false positive rate and false negative rate. If the testing data set is a very representative sample of the operation environment, we can use the metrics in the testing data to approximate the real world situation. In our experimental framework we assume all sensors are identical and configured to exhibit a detection rate of 95% and a false positive rate of 0.1%. These figures are in accordance with the features claimed by most IDS products.

### 3.2 Fitness Measurement

The fitness of a sensor placement is determined by its ability to satisfy three objectives: minimising the number of sensors, maximising detection rate and minimising false alarm rate.

Equation (1) is used to minimise the number of sensors, and the $nSensors$ represents the number of sensors.

$$f_1(P) = nSensors \tag{1}$$

Equation (2) is used to maximise the detection rate of a sensor placement. The $nDetectedAttacks$ represents the number of distinct attacks that have been detected; $nAttacks$ represents the number of all simulated attacks we have injected in the data set (i.e. 154 probe attacks). Note that we implement the sensors to detect attacks in a cooperative manner, which means duplication of alarms is avoided, and also cooperating sensors are able to detect attacks they may not detect independently.

$$f_2(P) = \frac{nDetectedAttacks}{nAttacks} \tag{2}$$

Equation (3) is used to minimise the false alarm rate of a sensor placement. The $nFalseAlarms$ represents the number of false alarms that are raised by the sensors. The $nAllAlarms$ represents the number of all alerts that are reported

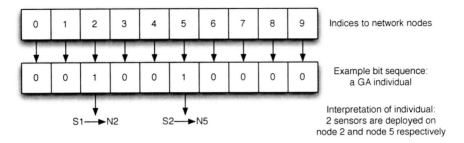

**Fig. 2.** Sensor Placement Representation

by the sensors. It is a sum of the number of detected attacks (a.k.a. true alarms) and the number of false alarms. So $f_3(P)$ follows precisely the definition of false alarm rate.

$$f_3(P) = \frac{nFalseAlarms}{nAllAlarms} \qquad (3)$$

### 3.3 Sensor Placement Representation

In our implementation, a feasible sensor placement is represented by $n$ (i.e. the number of network nodes) bits. Figure 2 is an example of how to interpret a bit sequence into a feasible sensor placement. In this example, we are going to deploy IDS sensors onto a small network of 10 nodes. There are 1023 (i.e. $2^{10} - 1$) distinct individuals, hence 1023 feasible sensor placements in total.

### 3.4 Parameters for the Search

Our implementation makes use of the versatile toolkit ECJ [9]. The major parameters for the GA search are as follows: the population size is 1500; the number of generations is 250; the crossover probability is 0.95 whereas the mutation probability is 0.05; the selection method is tournament of size 2.

To carry out multi-objective optimisation, an implementation of the Strength Pareto Evolutionary Algorithm 2 (SPEA2) algorithm was written as an extension to ECJ, which followed precisely the original algorithm specified by Zitzler et al [10]. The algorithm retains an archive of non-dominated individuals, which are individuals that cannot be improved upon in terms of all objectives by any other single individual within the archive. The algorithm attempts to use the archive to approximate the pareto front, a surface of non-dominated individual with objective space. We set the archive size for the multi-objective SPEA2 to 128.

The settings for parameters not listed here are given by the parameter files *simple.params* and *ec.params* supplied with the ECJ Toolkit. One of our experiment purposes is to demonstrate the validity and potential of the multi-objective approach to functional trade-offs in general, and so no parameter tuning was attempted.

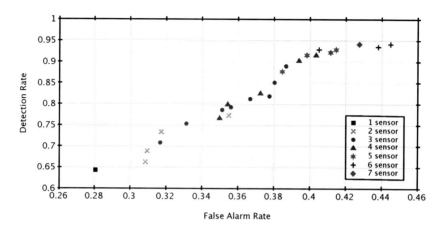

**Fig. 3.** Plot of Experiment Results

### 3.5 Experiment Results

We plot our experiment results in Figure 3, where each point corresponds to a placement's properties in the objective space[1]. The results validate our multi-objective optimisation approach and demonstrate that functional trade-offs are indeed possible for sensor placement problem.

Figure 3 shows the trend that the more sensors we use, the more attacks we will be able to detect (higher detection rates), whilst the more false alarms (higher false alarm rate) we will have to dismiss. Intuitively, by deploying multiple sensors on various network segments, we can tune each of them to the traffic that we typically see on that segment; due to the increased network visibility, more attacks are detected as more sensors are deployed. False alarm rate depends in practice on many factors (e.g. signature quality, volume of background traffic etc.). In this experiment, because we use sensors with the same settings, the false alarm rates were dominated by the volume of background traffic at different nodes. The more sensors are deployed, the higher volume of background traffic they will see, hence the higher false alarm rate.

Note that deploying more sensors may help to reduce false alarm rate in some situations. For example, both the placement with 7 sensors deployed on nodes 1, 12, 15, 16, 17, 18, 19 (the red diamond point on top right; also see Table 1) and the placement with 6 sensors deployed on nodes 3, 8, 9, 15, 16, 19 (the blue cross on top right) have a detection rate of 94.15%. However, the placement with 7 sensors has a lower false alarm rate of 42.75%. The false alarm rate of the placement with 6 sensors is 44.49%. This result means we may get better detection quality (in terms of the pair of detection rate and false alarm rate) with one more sensor.

---

[1] Note that this is not a ROC curve. The FA rate counts the fraction of FAs in the set of alerts generated, which is not equal to the false positive rate (fraction of non-attack events which raise an alarm).

**Table 1.** Example Placement Options of Varying Quality

| Sensors | Detection Rate | False Alarm Rate | Placement Options |
|---|---|---|---|
| 1 | 64.29% | 28.06% | Node 1 |
| 2 | 73.38% | 31.74% | Nodes 1, 3 |
| 2 | 77.27% | 35.48% | Nodes 3, 4 |
| 3 | 78.57% | 35.11% | Nodes 1, 3, 12 |
| 3 | 88.96% | 38.67% | Nodes 3, 5, 9 |
| 4 | 82.47% | 37.25% | Nodes 1, 3, 12, 19 |
| 4 | 91.56% | 40.34% | Nodes 3, 8, 9, 15 |
| 5 | 87.66% | 38.46% | Nodes 1, 3, 12, 18, 19 |
| 5 | 91.56% | 39.83% | Nodes 3, 4, 12, 15, 19 |
| 6 | 92.86% | 40.50% | Nodes 1, 3, 12, 15, 18, 19 |
| 6 | 94.16% | 44.49% | Nodes 3, 8, 9, 15, 16, 19 |
| 7 | 94.16% | 42.75% | Nodes 1, 12, 15, 16, 17, 18, 19 |

In the interests of brevity, we list some selected placements options that were determined in the experiments in Table 1. For example, the first sensor placed on node 1 is able to detect 64.29% of all attacks. This is due to the location of node 1 (see Figure 1). It provides a strategic advantage, as it serves to link over half of the network (through nodes 4 and 5) with the other half (through node 3).

## 4 Conclusions and Further Work

Means to reason and compare IDS sensor placements are important to judge the potential ability of such sensors to make a difference individually or in combination. The nature of sensor placement problem is such that there are too many criteria to consider when making a cost-effective decision, hence a multi-objective optimisation problem. Our experiments demonstrate the validity and potential of the multi-objective approach to sensor placement trade-offs and provide incremental placement options.

The work presented in this paper is a deliberate attempt to use GA and MOO techniques to assist network administrators to choose IDS sensor placement that effectively satisfies multiple criteria. The placement strategies generated, although simple, are typical places that network administrators would likely deploy IDS sensors. The ease with which the approach generated placements satisfying realistic security requirements merits further investigation of the technique. Experimentation and our general knowledge of intrusion detection systems have allowed us to identify numerous possible improvements to the approach and tool support. These are outlined below.

A straightforward extension of this work would be to incorporate an increased number of security requirements. Sensor placement is critical to providing effective defence. Optimal placement for this purpose would seek to minimise damage caused by intrusions. Placements that seek to maximise the number of victims detected could be useful in identifying locations best for detecting attacks likely to have more adverse impact. Such placements could be particularly important to detect and mitigate worm propagation and network probes (such as ping sweeps).

So far in the experiments, we did not consider the network having any other security measure deployed, as e.g. port access filters, firewalls, etc. This is indeed an important factor in a real scenario. We will take into account that information in our further work. Furthermore, we have dealt with network nodes in equal importance. In practice, some nodes are more significant to merit monitoring depending on the level of risk associated with individual nodes. Such level of risk needs to take into account both the value of assets and services offered and the likelihood of intrusions targeting them. One future work we are planning is to assign quantitative information (e.g. level of risk) to individual nodes and provide a model (e.g. the sensor deployment model by Shaikh [11]) to assess the information and incorporate it into the multi-objective optimisation framework.

## References

1. Goldberg, D.E.: Genetic Algorithms in Search, Optimization and Machine Learning. Addison-Wesley Longman Publishing Co., Inc., Boston (1989)
2. Coello, C.A.C., Nacional, L.: An updated survey of ga-based multiobjective optimization techniques. ACM Computing Surveys 32, 109–143 (1998)
3. Lu, W., Traore, I.: Detecting new forms of network intrusion using genetic programming. In: Proceedings of the 2003 Congress on Evolutionary Computation (2003)
4. Noel, S., Jajodia, S.: Attack graphs for sensor placement, alert prioritization, and attack response. In: Cyberspace Research Workshop (2007)
5. Rolando, M., Rossi, M., Sanarico, N., Mandrioli, D.: A formal approach to sensor placement and configuration in a network intrusion detection system. In: SESS 2006: Proceedings of the 2006 international workshop on Software engineering for secure systems, pp. 65–71. ACM, New York (2006)
6. Issariyakul, T., Hossain, E.: An Introduction to Network Simulator Ns2. Springer, Heidelberg (2008)
7. Shaikh, S.A., Chivers, H., Nobles, P., Clark, J.A., Chen, H.: Network reconnaissance. Network Security 11, 12–16 (2008)
8. Gu, G., Fogla, P., Dagon, D., Lee, W., Skoric, B.: Measuring intrusion detection capability: an information-theoretic approach. In: ASIACCS 2006: Proceedings of the 2006 ACM Symposium on Information, computer and communications security, pp. 90–101. ACM, New York (2006)
9. Luke, S.: A java-based evolutionary computation research system (2008), http://cs.gmu.edu/~eclab/projects/ecj/
10. Zitzler, E., Laumanns, M., Thiele, L.: Spea2: Improving the strength pareto evolutionary algorithm. Technical Report 103, Swiss Federal Institute of Technology (2001)
11. Shaikh, S.A., Chivers, H., Nobles, P., Clark, J.A., Chen, H.: A deployment value model for intrusion detection sensors. In: 3rd International Conference on Information Security and Assurance. LNCS, vol. 5576, pp. 250–259. Springer, Heidelberg (2009)

# Towards Ontology-Based Intelligent Model for Intrusion Detection and Prevention

Gustavo Isaza[1], Andrés Castillo[2], Manuel López[1], and Luis Castillo[3]

[1] Departamento de Sistemas e Informática, Universidad de Caldas, Calle 65 # 26-10, Manizales, Colombia
{gustavo.isaza,felipe}@ucaldas.edu.co
[2] Departamento de Lenguajes e Ingeniería del Software, Universidad Pontificia de Salamanca, Campus Madrid, Paseo Juan XXIII, 3, Madrid, Spain
andres.castillo@upsam.net
[3] Departamento de Ingeniería Industrial, Universidad Nacional de Colombia Sede Manizales, Colombia
lfcastilloos@unal.edu.co

**Abstract.** Nowadays new intelligent techniques have been used to improve the intrusion detection process in distributed environments. This paper presents an approach to define an ontology model for representing intrusion detection and prevention events as well as a hybrid intelligent system based on clustering and Artificial Neuronal Networks for classification and pattern recognition. We have specified attacks signatures, reaction rules, asserts, axioms using Ontology Web Language with Description Logic (OWL-DL) with event communication and correlation integrated on Multi-Agent Systems, incorporating supervised and unsupervised models and generating intelligent reasoning.

**Keywords:** Ontology, Intelligence Security, Intrusion Prevention, Multi-agent systems.

## 1 Introduction

The traditional taxonomies and the standard representations for Intrusion Detection Systems are insufficient to support the optimal attacks identification, definition or predicting similar anomalous traffic. An Intrusion Prevention System (IPS) is a security component that has the ability to detect attacks and prevent possible new abnormal traffic. Ontologies allow describing objects, concepts, and relationships in a knowledge domain, therefore resources as signatures, reaction and prevention rules need to be sufficiently described to be published, accessed, invoked and reasoned. Having an ontological model gives to our agent platform a solid structure to improve knowledge representation, providing a homogeneous communication and integrating an intelligent reasoning to optimize functionality the intrusion detection tasks. Ontology can supply intrusion detection and prevention features giving the ability to share a common conceptual understanding and providing relationships between heterogeneous components.

This paper aims to present a progress in a researching project using an ontological model to represent intrusion detection and prevention over multi-agents architectures and using intelligence computing for reasoning, classification and pattern recognition. The remainder of this paper is organized as follows: The section 2 discusses the relevant previous work in this area, ontology used for attacks signatures and reaction rules to support the prevention system is presented in section 3 and the integration with our multi-agent system, in section 4 we present the intelligence technique applied with Artificial Neuronal Networks and a clustering technique using K-Means. Finally we summarize our research work and discuss the future contributions.

## 2 Previous Work

The research developed in [1] has defined a target centric ontology for intrusion detection, new approaches have proposed ontologies as a way to represent and understand the attacks domain knowledge, expressing the intrusion detection system much more in terms of their domain and performing intelligent reasoning. These projects aim to define an ontology DAML-OIL (DARPA Agent Markup Language + Ontology Interface) target centric based on the traditional taxonomy classification migrated to semantic model, the investigation done in [2] integrates ontology entities and interactions captured by means of an centric-attack ontology which provides agents with a common interpretation of the environment signatures which are matched through a data structure based on the internals of the Snort network intrusion detection tool. A Security Ontology for DAML+OIL in order to control access and data integrity to Web resources was developed in [3]. Another projects have integrated multi-agent systems (MAS) in the intrusion detection problem (IDP) [4], [5]. The use of intelligence computing in the intrusion detection problem has been emphasized in supervised and non supervised models applying Neural Networks (ANN) [6], [7], to optimize the pattern recognition and novel attacks. Several mechanisms for classification attacks have been proposed such SVM (Support Vector Machines) [8], Data mining [9], and clustering [10], [11]. Our work suggests not only the integration of intelligent techniques for pattern recognition, besides the definition for intrusion events and prevention rules in the ontological model.

## 3 Ontology and Semantic Model

The Intrusion Detection Messages Exchange Format (IDMEF) [12] was created to enable interoperability between heterogeneous IDS (Intrusion Detection Systems) nodes, representing a specification for a standard alert format and a standard transport protocol. IDMEF use a hierarchal configuration with three main elements (sensors, managers and analyzers). The main disadvantage using IDMEF is the use of XML that it becomes a syntactic representation and does not allow generation of reasoning or inference [1]. The OWL is a descriptive language; its design has multiple influences from established formalisms and knowledge representation paradigms, derived from existing ontology languages and based on existing Semantic Web standards [13], the specification was based in description logics and the RDF/XML exchange syntax. Description Logics (DLs) have relevant influences on the OWL design, especially on the formalisation semantics, the data-type integration and their knowledge representation (KR).

The problem presented in our model defines multiple concepts, relations and the meaning integration with other security technologies like firewalls and vulnerabilities scanner. Ontologies can help to optimize the knowledge representation, reduce the messages size, decrease the redundancy and allow incorporating more intelligence in the information analysis. The Intrusion data definition and ontology model, their design processes and life cycle have been represented following *Methontology* [14] to describe the concepts, relations, asserts, axioms, rules and instances. An example of a formal axiom and a Rule representation is shown in Table 1. To define the rules we are using the Semantic Web Rule Language (SWRL), it is an expressive OWL-based rule language. SWRL allows writing rules expressed in OWL concepts providing reasoning capabilities. A SWRL rule contains an antecedent part to describe the *body*, and a consequent part, referring the *head*. The body and head are positives conjunctions of *atoms*.

**Table 1.** Formal Axiom and Rule Table

| Name | Expression |
|---|---|
| RootAccess Axiom Definition | RootAccess ≡ ∃ (IntrusionState ∩ InputTraffic ∩ NetworkNode) ∩ ∃ Generated_by(AnomalousSSH ∪ AnomalousFTP ∪ WebAttack ∪ TelnetAccess) ∩ UID_Try(UID_0) |
| WebAccess Attack Rule | Network_Node(?nr) ^ Input_Traffic(?it) ^ Intrusion_State(?is) ^ WebInjection(?wi) ^ Generated_by(?wi, ?is) -> NodeUnderWebInjection(?nr) |

At the moment we have defined around 1800 attacks signatures and 835 prevention rules instances in our ontology, in addition we described the events sequence that happen in real time. The Reaction Agent generates the Intrusion Prevention events sending rules messages to other network components reconfiguring dynamically the security environment. The predicates definition describes a sequence attack having input causes and possible consequences. The input data was generated using DARPA Data Sets Intrusion Detection Evaluation [15], additionally the attacks were created with testing tools as *Metasploit*, *IDSWakeUP*, *Mucus* synchronized with Snort, normal traffic captured using *JPCap* and *tcpdump* and processed in the multi-agent for IDS and IPS (*OntoIDPSMA* – Ontological Intrusion Detection and Prevention Multi-agent system). The raw data was converted to XML then processed to OWL instances; furthermore the ontology is updated via SPARQL sentences. The ontology represents the signatures for known attacks and novel attacks, the intelligent behaviour uses the inference model and reasoning tools integrating neuronal networks in the multi-agent system; this implementation provides an ontological model for reaction rules creating the prevention system. A fragment of the ontology defined that implements the intrusion detection and prevention knowledge is depicted in Fig. 1 that presents a high level view.

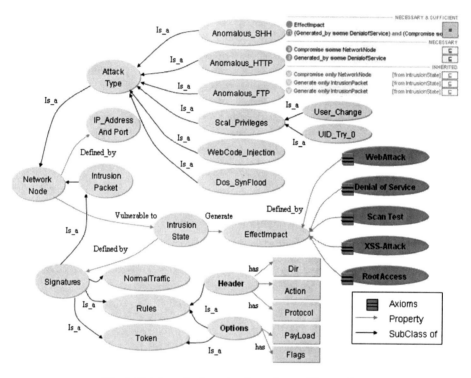

Fig. 1. Ontology for Intrusion Detection and Prevention

The behaviour for the Intrusion Detection and Prevention multi-agent has been designed using a semantic model integrated with the signature and reaction rule ontology. Fig. 2 shows this integration incorporating the supervised and unsupervised intelligence technique in selection and pattern recognition phase.

## 4 Classifier and Pattern Recognition Model

The intelligence technique has been implemented using a supervised model applying Neuronal Networks; multiple relevant fields were captured, normalized, processed and classified (Header IP Addresses, Port information, Priority, Protocol, Data, Type of Service among others). We have described this implementation in [16]. The Sum Squared Error as function of epoch number in the learning process is depicted in Fig. 3.

At the moment we are probing a clustering and classification model using K-means, this is an unsupervised learning algorithm that solves clustering problems. This technique tends to minimize an objective function, the squared error function. The objective function:

$$J = \sum_{j=1}^{K} \sum_{i=1}^{n} \left\| x_i^j - c_j \right\|^2. \tag{1}$$

Where $\left\| x_i^j - c_j \right\|^2$ is a distance between data represented in $x_i^j$, the cluster centre $c_j$ indicates the distance of the *n* data points from their own cluster centres.

The data gathered via *JPCap* and the KDD Intrusion Detection Datasets, are classified, then, the relevant fields were selected and clustered. For now the model defines the cluster number *k* of each class, and the number of samples to be extracted.

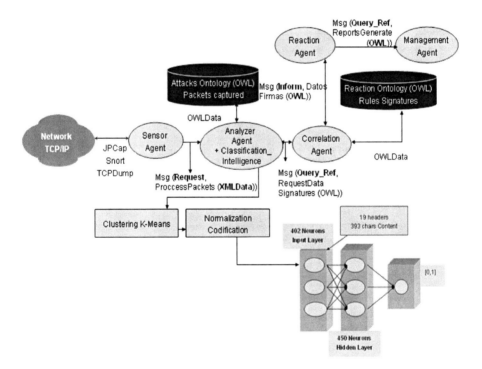

**Fig. 2.** *OntoIDPSMA* Multi-Agent System Architecture

We probed multiple classifiers to evaluate our Dataset, to compare; we have registered the True Positives, False Positives and False Negatives for each algorithm as well as the elapsed and training time.

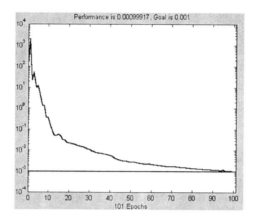

**Fig. 3.** Training Error Learning Process. Sum Squared Error vs. Epoch Number.

The results achieved using the best performance configurations are presented in Table 2.

**Table 2.** Model Behaviour using the classification and the intelligence techniques

| Packets | Packets | False Positives | False negatives | Uncertainty Packets | % traffic successful | Elapsed Time |
|---|---|---|---|---|---|---|
| Training | 1955 | 0 | 1 | 0 | | |
| Validation | 101 | 0 | 0 | 0 | | |
| Normal | 50000 | 752 | - | 573 | 97,94% | 285 sec |
| Anomalous | 246 | - | 17 | 4 | 99,05% | |

The complete architecture performance integrating the Ontologies in our Multi-Agent System and incorporating the intelligence behaviour is shown in Table 3.

**Table 3.** Performance using MAS for Intrusion Detection and Prevention

| Data Analyzed | Standard IDS | OntoIDPSMA |
|---|---|---|
| Traffic Processed | 94,4 % | 93,72 % |
| Packets Processed | 51314 | 51093 |
| Anomalous packets successful (detected) | 92,3 % | 99,06% |

The Fig. 4 shows the clustering results using as the main attribute the *payload* (Data) and other (IP, Port Source and Destination, Protocol, Type Of Service, Flags); the first graphic displays 4 clusters clearly identifying the groups used in our samples with the attacks events detected with our ontology rules, the second shows 3 clusters identifying the attacks group in the unsupervised model. We will continue testing new

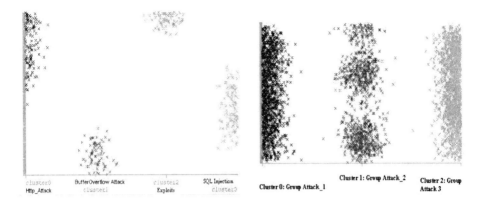

**Fig. 4.** Cluster for Data Payload Classification

configurations to achieve an optimized classification results and to improve the clustering and an approach pattern classification integrated with the Neuronal Network architecture.

## 5 Conclusions and Future Work

As a result, Ontology integrated with Multi-Agent System in the Intrusion Detection and Prevention problem has been developed and ready to be probed using inferences and reasoning models. The prevention architecture based on reaction rules generated by the intelligent and correlated component in our OntoIDPSMA creates new reaction rules in other security components. This paper has presented an ontological model integrated with multi-agent system, using intelligence computing providing better performance in processes such as knowledge representation, cooperation, distribution, intelligence reasoning, reactivity, among others; the prediction capabilities have been implemented using Neuronal Networks and the clustering process is being developed and tested using K-Means. In Table 2 and 3 we presented the training results and the performance testing applying our architecture and traditional IDS. So far, our project is in development, the whole system is being implemented and integrated. In the future, we will work on improvements the analyzer agent, integrating hybrid intelligence techniques that allow reach better results.

## References

1. Undercoffer, J., Finin, T., Joshi, A., Pinkston, J.: A target centric ontology for intrusion detection: using DAML+OIL to classify intrusive behaviors. In: Knowledge Engineering Review - Special Issue on Ontologies for Distributed Systems, pp. 2–22. Cambridge University Press, Cambridge (2005)
2. Mandujano, S., Galvan, A., Nolazco, J.: An ontology-based multiagent approach to outbound intrusion detection. In: The 3rd ACS/IEEE International Conference on Computer Systems and Applications, p. 94 (2005)

3. Denker, G., Kagal, L., Finin, T.W., Paolucci, M., Sycara, K.: Security for DAML web services: Annotation and matchmaking. In: Fensel, D., Sycara, K., Mylopoulos, J. (eds.) ISWC 2003. LNCS, vol. 2870, pp. 335–350. Springer, Heidelberg (2003)
4. Dasgupta, D., Gonzalez, F., Yallapu, K., Gomez, J., et al.: CIDS: An agent-based intrusion detection system. Computer and Security: Science Direct 24(5), 387–398 (2005)
5. Herrero, A., Corchado, E., Pellicer, M., Abraham, A.: Hybrid Multi Agent-Neural Network Intrusion Detection with Mobile Visualization in Innovations in Hybrid Intelligent Systems, pp. 320–328. Springer, Heidelberg (2008)
6. Golovko, V., Kachurka, P., Vaitsekhovich, L.: Neural Network Ensembles for Intrusion Detection. In: 4th IEEE Workshop on Intelligent Data Acquisition and Advanced Computing Systems: Technology and Applications. IDAACS 2007, pp. 578–583 (2007)
7. Laskov, P., Dussel, P., Schafer, C., Rieck, K.: Learning intrusion detection: Supervised or unsupervised? In: 13th International Conference on Image Analysis and Processing - ICIAP, Cagliari, Italy, pp. 50–57 (2005)
8. Li, K., Teng, G.: Unsupervised SVM Based on p-kernels for Anomaly Detection. In: Proceedings of the First International Conference on Innovative Computing, Information and Control, vol. 2. IEEE Computer Society, Los Alamitos (2006)
9. Zurutuza, U., Uribeetxeberria, R., Azketa, E., Gil, G., et al.: Combined Data Mining Approach for Intrusion Detection. In: International Conference on Security and Criptography, Barcelona, Spain (2008)
10. Al-Mamory, S., Zhang, H.: Intrusion detection alarms reduction using root cause analysis and clustering, pp. 419–430. Butterworth-Heinemann (2009)
11. Jiang, S., Song, X., Wang, H., Han, J., et al.: A clustering-based method for unsupervised intrusion detections, pp. 802–810. Elsevier Science Inc., Amsterdam (2006)
12. IETF-IDMEF. he Intrusion Detection Message Exchange Format (IDMEF). Consulted (2008), http://www.ietf.org/rfc/rfc4765.txt (2007)
13. Horrocks, I., Patel-Schneider, P., McGuinness, D.: OWL: a Description Logic Based Ontology Language for the Semantic Web. In: Baader, F., Calvanese, D., McGuinness, D.L., Nardi, D., Patel-Schneider, P.F. (eds.) The Description Logic Handbook: Theory, Implementation and Applications, 2nd edn., pp. 458–486. Cambridge University Press, Cambridge (2007)
14. Corcho, Ó., Fernández-López, M., Gómez-Pérez, A., López-Cima, A.: Building legal ontologies with METHONTOLOGY and webODE. In: Benjamins, V.R., Casanovas, P., Breuker, J., Gangemi, A. (eds.) Law and the Semantic Web. LNCS (LNAI), vol. 3369, pp. 142–157. Springer, Heidelberg (2005)
15. DARPA. DARPA Intrusion Detection Evaluation, The, DARPA off-line intrusion detection evaluation. LINCOLN LABORATORY Massachusetts Institute of Technology. Consulted (2008),
http://www.ll.mit.edu/IST/ideval/data/1999/
1999_data_index.html (1999)
16. Isaza, G., Castillo, A., Duque, N.: An Intrusion Detection and Prevention Model Based on Intelligent Multi-Agent Systems, Signatures and Reaction Rules Ontologies in Advances in Intelligence and Soft Computing. In: Demazeau, Y., et al. (eds.) PAAMS 2009, pp. 237–245. Springer, Heidelberg (2009)

# Ontology-Based Policy Translation

Cataldo Basile, Antonio Lioy, Salvatore Scozzi, and Marco Vallini

Politecnico di Torino
Dip. di Automatica ed Informatica
Torino, Italy
{cataldo.basile,antonio.lioy,marco.vallini}@polito.it,
salvatore.scozzi@gmail.com

**Abstract.** Quite often attacks are enabled by mis-configurations generated by human errors. Policy-based network management has been proposed to cope with this problem: goals are expressed as high-level rules that are then translated into low-level configurations for network devices. While the concept is clear, there is a lack of tools supporting this strategy. We propose an ontology-based policy translation approach that mimics the behaviour of expert administrators, without their mistakes. We use ontologies to represent the domain knowledge and then perform reasonings (based on best practice rules) to create the configurations for network-level security controls (e.g., firewall and secure channels). If some information is missing from the ontology, the administrator is guided to provide the missing data. The configurations generated by our approach are represented in a vendor-independent format and therefore can be used with several real devices.

## 1 Introduction

One major weakness in today security landscape is still an old one: the human factor. Already back in the year 2000, the "Roadmap for defeating Denial of Service attacks"[1] highlighted the general *scarce technical talent* and *decreased competence* of system administrators. Unfortunately, this scenario has not significantly improved, as confirmed by recent reports [1]. It is therefore important to support security managers with automatic tools to minimize human errors.

Presently, policy-based network management (PBNM) [2,3] seems the best approach to cope with system administration because it separates the goals (i.e., the policy) from the mechanisms to achieve them (i.e., the security controls). Policies are expressed as high-level security statements, derived from business goals or best-practice rules. However actual security controls are mostly placed at network level. Firewall and virtual private network (VPN) are ubiquitous in security architectures because they have excellent performance (compared to application-level controls) and do not require changes to business applications. Evidence says that, as the network grows bigger, the configuration of network-level controls becomes exponentially complex. We have therefore a mismatch: policies are expressed at high level while controls are at network level. Bridging this gap is the task of *policy translation* that consists in transforming a policy from the level of abstraction A to the (lower) level B. This procedure is repeated until

---
[1] http://www.sans.org/dosstep/roadmap.php

the policy is expressed in a way that can be used to configure a specific control (e.g., IP-level filters for a firewall).

Quite often translation from high-level policies to actual controls is manually performed by security administrators but this is still time-consuming and error prone: if policy translation is not automated, PBNM is simply a problem shift, not a solution. Our approach, called *Ontology-Based Policy Translator* (OPoT), handles this problem using ontology-based reasoning to refine policies into configuration for the actual controls. In particular, OPoT mimics skilled administrators in collecting all the needed information and applying best practice for security configuration. Ontologies are very effective in capturing, defining, sharing, and reusing the knowledge about a specific domain. In particular, they are good at representing relationships between entities and at verifying the correctness of knowledge [4]. Therefore we use an ontology to represent the domain of interest – computer networks – from the topological and functional point of view, and the environmental data necessary to correctly configure them.

OPoT exploits ontology-based reasoning to understand the information necessary for policy translation and to enrich the knowledge base in order to fill the gap between abstraction levels. In other words, automatic reasoning is used to enlarge the domain of knowledge of the ontology without the need to involve human administrators. Nevertheless, the reasoning alone is unable to fully translate a policy because some information cannot be guessed or derived (e.g., the users' roles). To this purpose, OPoT guides the administrator in providing the data necessary to complete the configurations. OPoT also reports about possible anomalies when it discovers that best practice is not followed.

The advantages of this approach are manifold. First, it addresses policy translation in a way that reduces the impact of administrators' mistakes in the configuration workflow: we just assume that the provided information (e.g., organization charts, tasks) is correct. Second, it strongly decreases the effort and the time to have a set of correct configurations (from days to minutes). Finally, it quickly adapts to changes in the methodologies. The proposed approach can still be improved in several aspects, but our initial experimental results demonstrate its power as well as its extensibility.

## 2 Background and Related Work

The adoption of security ontologies is continuously increasing for network security and risk management fields. Strassner suggests that the traditional information and data models are not capable to describe detailed semantics required to reason about behaviour. His work [5] modifies the existing DEN-ng policy model to support and generate ontologies for governing behaviour of network devices and services.

Tsoumas et al. [6] develop a security ontology to perform network security management thus modelling assets (e.g., data, network, services), countermeasures (e.g., firewall, antivirus) and the related risk assessment. They also describe a framework for security knowledge acquisition and management.

Fenz et al. [7] define a knowledge model (defining a security ontology) to support risk management domain, incorporating security concepts like threats, vulnerabilities, assets, controls, and related implementation. . Another work [8] proposes an ontology-based approach to model risk and dependability taxonomies. The related framework

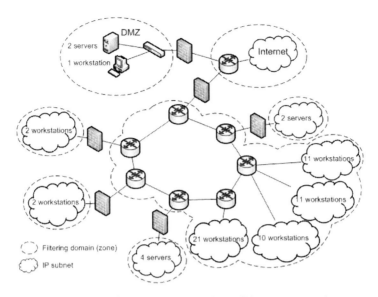

**Fig. 1.** The reference network used to validate our approach

contains a tool for simulating threats against the modeled ontology. KAoS represents another approach to policy-based management using the description logic and ontologies from representation to conflict analysis and translation [9,10].

## 3  Case Study

We present in Fig. 1 the network use as a case study to validate our approach and to present its main results and implementation. The network contains 90 nodes divided into the following categories: workstation, server, router, firewall and switch. A server is a computer that provides a service through a Service Access Point (SAP). The network is organized into different IP subnetworks, according to different company's departments and it includes also a DMZ (De-Militarized Zone). The network is divided into seven filtering domains, called zones [11]. Each zone is delimited by network elements with filtering capability (firewalls in this case study).

In this network we will configure two categories of security controls: traffic filtering and channel protection policies. Enforcing these kind of policies is theoretically easy, nevertheless they are constant sources of problems and mistakes, especially when the size on the network grows and for non-experienced administrators [12].

## 4  Our Approach

The objective of our ontological framework is to derive configurations for security controls - based on ACL (Access Control List) and secure channel mechanisms - from a fixed set of business-oriented policies using automatic reasonings and interactions with the security administrator. The main phases of this process are presented in Fig. 2. The

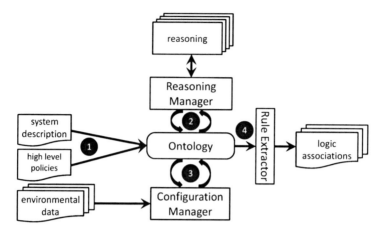

**Fig. 2.** The logical workflow of the OPoT system

security administrator is the person in charge of configuring network security: we assume that he can access all the information required to generate configurations.

The ACL controls are derived both for devices (e.g., router, managed switch, firewall) and end-nodes. On the end-nodes we can distinguish two classes of controls: OS-level (e.g., local firewall) and application-level (e.g., web server access control). We also aim at configuring secure channels: TLS (end-to-end) and IPsec (end-to-end, gateway-to-gateway and end-node-to-gateway). The output of this process is represented in a vendor-independent format inspired to the CIM Simplified Policy Language [13] and IETF Policy Core Information Model [14] and its extensions [15].

To simplify the problem, the configurations are obtained as translation of a set of twelve basic policies. Every policy is a statement describing a business-level security objective. Examples are "every user must access only the services he is authorized to use", "business services must be connected securely to their components located in other parts of the network at higher security". We analyzed different sources to decide the sample policies to support. First we combined our experience and interviews with the security administrators in our institution (representative of an open environment with several unrelated actors, multiple levels of security and untrusted users) and partners (representative of private networks handling highly sensitive data and hosting some public services). Moreover, we analyzed different policy models (e.g., Role Based Access Control (RBAC) [16] and Mandatory Access Control [17]). Finally, we examined publicly available policy templates, in particular the SANS Security Policy Project [18].

Together with policies, OPoT takes as input the initial security ontology containing the class hierarchy and the system description represented in the P-SDL language [19].

The process of policy translation as knowledge refinement and enrichment has been designed to mimic the behaviour of skilled and experienced administrators in acquiring environmental information and translating it to actual rules to enforce best security practices. Knowledge enrichment is performed through two entities, the *Reasoning Manager* (RM) and the *Configuration Manager* (CM). The RM coordinates a set of *reasonings* that apply logical inference on the knowledge base to increment it using a standard ontology reasoner. Example of reasonings are the automatic identification of DMZs, the classification of services according to their scope, the classification of SAP. However, not all the required information can be derived from the inputs. The CM coordinates the import of external/environmental non-deducible information by driving the administrator to collect missing information. For example, OPoT may asked to provide the host-user assignments, the user's task assignment according to the company's organization charts (e.g., expressed using RBAC), or description of business services provided through the network (e.g., the WS-CDL description of services).

The logical workflow of our framework is simple. First of all, the administrator chooses among the available policies. Every policy entails the type of information required to refine it, that is, the reasoning and data to be acquired to be enforced. Then, system description is used to populate the initial ontology. Subsequently, OPoT asks RM to execute the implied reasonings. Reasonings also warn the administrator about possible anomalies. If needed, CM asks the administrator to provide missing data through a graphical user interface. RM runs always before CM because it does not require effort from the administrators. This cycle may run more than once until the policy is enforceable. If in a cycle the ontology is not modified and the goals are still not achieved, the policy is considered not enforceable and a warning message is sent to the administrator.

The last entity, the *Extractor*, derives the logical associations from the ontology and then it completes the translation. A logical association expresses an interaction between parties (typically one-to-many) and the related communication properties (filtering constraints, e.g., need-to-access condition). But it is not our desired output, yet. The translation process distinguishes between topological-independent and topological-dependent logical associations. In the first case (e.g., end-to-end interactions such as TLS protected channel) the process directly generates the configurations for the end nodes. In the second case, the network topology needs to be analyzed. All paths between the source and destination nodes are identified and rules are generated for the involved devices. For example, when a company client need to reach an internal service, OPoT configures all the firewalls in the paths between them.

## 4.1 The Security Ontology

Our security ontology is structured in three levels (Fig. 3). The first one contains classes whose instances cannot be derived from other classes. First level instances are created gathering information contained in external files and running algorithms to manipulate it (e.g., filtering zones). The next levels are generated through either the usage of a standard reasoner (in our case Pellet [20]) or using the reasonings. The higher the level, the more detailed the information; for example, in the second level computers are classified in workstations or servers and in the third level the workstations are further classified in shared or personal. The security ontology contains several properties linking the instances of different classes. This permits to navigate and analyze the ontology for deduction purposes. Thus, for example, an instance in the server class has a property that permits linking it to the instances of the SAP and it is linked to the Ciphersuites.

The ontology has been defined to perform a first automatic classification of second/third level classes. For instance in Fig. 4 is presented the OWL class definition that

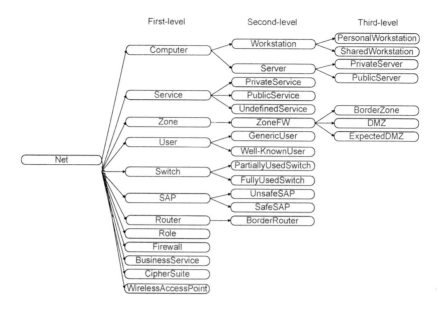

**Fig. 3.** Security ontology

classifies zones as DMZ if they contains public server and they are linked to a border zone (zone including an access to Internet). If a zone satisfies only the first condition, it is considered as expected DMZ and it will be analyzed to check its characteristics. After system description is scanned, a reasoner is run to perform the automatic classification.

Every reasoning is characterized by a set of input ontology classes and a set of output ontology classes, subclasses of the input ones. The reasoner tries to "move" the instances of input classes towards higher levels of details, that is, to output classes. A set of SPARQL [21] queries permits to retrieve the instances to be analyzed together with all the data useful for the decision.

For instance, the ServerReasoning takes as input the instances of Service classes, tries to understand if they provide functionalities that should be visible from the outside. In order to do this, the reasoning collects information about the firewall zones and possibly the potential DMZ, if already identified. For example, a service that is reachable from the Internet without encountering a firewall is considered public, but the administrator is warned about a possible anomaly.

The CM is organized in a set of modules, each one is able to insert information into the ontology. For this reason each configurator module is characterized by the classes whose instances may be inserted, the classes whose instances can be modified and the properties that can be added. For instance, the UsersRoleConfigurator module manages the RBAC information and it only inserts instances of the RBAC classes while the ServiceConfigurator, able to parse WS-CDL information, may move instances of the Service classes to PublicService or PrivateService classes and it can add the properties NeedtoAccess, a property we introduced to represent the network connections that must be guaranteed for the correct policy enforcement. Additionally, both the configurator modules and reasonings contain precedences, a list of modules that must be run before.

```xml
<owl:Class rdf:about="#DMZ'>
    <owl:equivalentClass>
        <owl:Class>
            <owl:intersectionOf rdf:parseType="Collection">
                <owl:Restriction>
                    <owl:onProperty rdf:resource="#contains"/>
                    <owl:someValuesFrom rdf:resource="#PublicServer"/>
                </owl:Restriction>
                <owl:Restriction>
                    <owl:onProperty rdf:resource="#isLinkedTo"/>
                    <owl:someValuesFrom rdf:resource="#BorderZone"/>
                </owl:Restriction>
            </owl:intersectionOf>
        </owl:Class>
    </owl:equivalentClass>
    <rdfs:subClassOf rdf:resource="#ZoneFW"/>
</owl:Class>
<owl:Class rdf:about="#ExpectedDMZ">
    <owl:equivalentClass>
        <owl:Restriction>
            <owl:onProperty rdf:resource="#contains"/>
            <owl:someValuesFrom rdf:resource="#PublicServer"/>
        </owl:Restriction>
    </owl:equivalentClass>
    <rdfs:subClassOf rdf:resource="#ZoneFW"/>
</owl:Class>
<owl:Class rdf:about="#PublicServer">
    <owl:equivalentClass>
        <owl:Restriction>
            <owl:onProperty rdf:resource="#provides"/>
            <owl:someValuesFrom rdf:resource="#PublicService"/>
        </owl:Restriction>
    </owl:equivalentClass>
    <rdfs:subClassOf rdf:resource="#Server"/>
</owl:Class>
<owl:Class rdf:about="#PublicService">
    <rdfs:subClassOf rdf:resource="#Service"/>
</owl:Class>
```

**Fig. 4.** An extract of the OWL ontology that tries to automatically classify the DMZ

Every policy is characterized by its "goal", that is, the properties that must be derived to implement it, and a set of subsidiary classes that indicate from which information the properties must be derived. In the next section will be presented the case of the "Every user must access only the services he is authorized to use" policy, having as goal the NeedToAccess property and auxiliary classes the User class and the Service class.

The OPoT tools uses information about the single components to decide the ones to execute and the order. Starting from a policy, the tool tries to identify the modules able to derive the goal properties of the policy, then it identifies the dependencies, e.g., which components involve classes that are used as input for detected modules. The result is a dependency graph, later processed by RM and CM. This approach allows the maximum flexibility. In fact, every time a new module is added it is enough to specify which classes it uses and OPoT can automatically decide when they can be used. Nevertheless, a convergence problem arises, indeed, modules may work on the same data originating a loop in the dependency graph. In our example, the convergence is assured because existing modules have been designed to reduce it and because the entire cycle is repeated until the policy is correctly implemented. The solution of this problem will be improved in future versions of the tool.

### 4.2 An Example of Policy Translation

As an example we present here the case of derivation of filtering rules associated to the policy "Every user must access only the services he is authorized to use". This represents a typical goal in many organizations, implicitly adopted during network configuration. Translating this policy means, as a minimum, to insert all the necessary NeedToAccess properties inside the ontology from components that manage the Service class and the User class. The policy codes in term of ontology classes the best practice stating that a user is supposed to be authorized to access all the private services he needs to use for performing his work, all the company's public services and the Internet, according to the "Remote Access Policy" [18]. A particular attention is devoted to shared workstation. They are nodes usable by several types of users and therefore must be able to access all the services needed by these users to perform their work.

OPoT must define the NeedToAccess associations, thus it "understands" that the first component to run is the UsersRoleConfigurator. In fact, in principle it is not possible to deduce which services a user needs to use, the CM asks administrator to include the explicit user-role RBAC description also containing the remote access policy.

OPoT maps users and services to IP nodes. Service information are present in the ontology in form of SAPs obtained from the system description. The association between users and IP addresses is a typical information that cannot be derived by the system description. For this reason, OPoT asks the administrator to provide the workstation mapping: a user is assumed to use his workstation or a shared one.

OPoT then deduces from auxiliary classes as well as input classes that services must be classified. As a pre-requisite the DMZ must be identified. The objective of a DMZ is to provide services to a zone considered at a lower security level. Usually a public DMZ is a subnetwork directly connected to the border firewall (if there is one) or to a firewall connected to a border router where services are located. The Internet is supposed at security level 0, while the internal network and the DMZ are at level 1. Then, the reasoning looks for other internal DMZs, it identifies the separation domains, and assigns security levels accordingly. The assumption is that a service in DMZ at security level $l$ must be available from zones having security level $l-1$ but not from the ones at level $l-2$. For services for which it is not possible to identify the scope, administrator is asked to provide the classification. This reasoning reports a set of anomalies to the administrators: the DMZ should not contain workstations and services in zones at security level $l$ should not be accessible to hosts in zones at level $l-2$ or smaller creating breaches in the security "compartments". After CM and RM have run all the identified components and possibly repeated the workflow, all the needed properties are derived.

## 5 Implementation

OPoT is implemented in Java (Fig.5) using the Jena API [22] to manage the security ontology. The system description and the environmental data are expressed in XML using standards when available (e.g., RBAC profile of XACML [23]). Our security

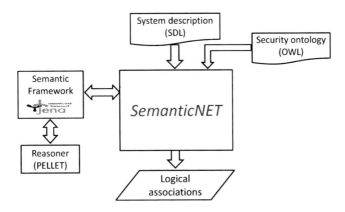

**Fig. 5.** Architecture

ontology is written in OWL-DL [24] and the software used to edit it is Protégé [25]. A set of classes parses the system description and runs a series of graph-based algorithms to extract additional information (e.g., the filtering zones). The policies are represented as Java classes. A policy links the reasonings to gather the information needed to its translation. Also reasonings are implemented as Java classes that interact with the ontology through the Jena framework and the Pellet reasoner [20] using the Description Logic Interface (DIG) [26].

While deriving the logical associations, OPoT performs several controls, according to best practice, to find ambiguities or conflicts. In case of anomalies, it shows to administrator the list of the problems and suggests him the possible solutions. Finally, the logical associations are extracted using SPARQL [21] to query ontology.

Considering the network in Fig.1 and seven out of twelve policies, the tool spends about two minutes of CPU time from the policy selection to the translation. This test was performed using a PC with a 2 GHz CPU and 1 GB of RAM. The max memory utilization was about 30 MB.

## 6 Conclusion and Future Work

This paper presented OPoT, an ontology-based approach to policy translation. It uses a security ontology to drive the administrator from high-level policy specification downto system configuration, mainlt at IP level. In principle, OPoT can be extended to use other network security controls, such as IEEE 802.1x. Moreover, automatic reasoning is currently able to cope only with a pre-defined set of policies. In future, when the automatic semantic analysis of text will came of age, we hope to derive the policies directly from a high-level textual specification.

## References

1. Agrawal, D.: Business impact of research on policy for distributed systems and networks. In: IEEE POLICY 2007, Bologna, Italy (June 2007)
2. Westerinen, A., Schnizlein, J., Strassner, J., et al.: Terminology for Policy-Based Management. RFC-3198 (November 2001)

3. Strassner, J.C.: Policy Based Network Management. Morgan Kauffman Publishers, San Francisco (2004)
4. Gruber, T.R.: Toward Principles for the Design of Ontologies Used for Knowledge Sharing. Int. Journal Human-Computer Studies 43(5-6), 907–928 (1995)
5. Strassner, J., Neuman de Souza, J., Raymer, D., Samudrala, S., Davy, S., Barrett, K.: The design of a new policy model to support ontology-driven reasoning for autonomic networking. In: LANOMS 2007, Rio de Janeiro, Brasil, September 2007, pp. 114–125 (2007)
6. Tsoumas, B., Gritzalis, D.: Towards an ontology-based security management. In: Int. Conf. on Advanced Information Networking and Applications, Vienna, Austria, pp. 985–992 (2006)
7. Fenz, S., Ekelhart, A.: Formalizing information security knowledge. In: ASIACCS, Sydney, Australia, pp. 183–194 (2009)
8. Ekelhart, A., Fenz, S., Klemen, M., Weippl, E.: Security ontologies: Improving quantitative risk analysis. In: Hawaii Int. Conf. on System Sciences, Big Island, Hawaii, p. 156a (2007)
9. Uszok, A., Bradshaw, J.M., Johnson, M., Jeffers, R., Tate, A., Dalton, J., Aitken, S.: KAoS policy management for semantic web services. IEEE Intelligent Systems 19(4), 32–41 (2004)
10. Uszok, A., Bradshaw, J., Lott, J.: et al.: New developments in ontology-based policy management: Increasing the practicality and comprehensiveness of KAoS. In: IEEE POLICY 2008, Palisades, NY, USA, June 2008, pp. 145–152 (2008)
11. Mayer, A., Wool, A., Ziskind, E.: Offline firewall analysis. Int. J. Inf. Secur. 5(3), 125–144 (2006)
12. Al-Shaer, E., Hamed, H.: Modeling and Management of Firewall Policies. IEEE Transactions on Network and Service Management 1(1), 2–10 (2004)
13. Agrawal, D., Calo, S., Lee, K.W., Lobo, J.: Issues in designing a policy language for distributed management of it infrastructures. In: IFIP/IEEE Int. Symp. on Integrated Network Management, Munich, Germany, pp. 30–39 (2007)
14. Moore, B., Ellesson, E., Strassner, J., Westerinen, A.: Policy core information model (RFC-3060) (February 2001)
15. Moore, B.: Policy core information model (PCIM) extensions (RFC-3460) (January 2003)
16. NIST: Role based access control, http://csrc.nist.gov/groups/SNS/rbac/
17. Loscocco, P.A., Smalley, S.D., Muckelbauer, P.A., Taylor, R.C., Turner, S.J., Farrell, J.F.: The inevitability of failure: The flawed assumption of security in modern computing environments. In: National Information Systems Security Conf., Crystal City, VA, USA, pp. 303–314 (1998)
18. SANS: The SANS Security Policy Project, http://www.sans.org/resources/policies/
19. POSITIF Consortium: The POSITIF system description language (P-SDL) (2007), http://www.positif.org/
20. Clark, Parsia: Pellet: The open source OWL DL reasoner, http://clarkparsia.com/pellet
21. Clark, K.G., Feigenbaum, L., Torres, E.: SPARQL protocol for RDF, http://www.w3.org/TR/rdf-sparql-protocol/
22. HP-Labs: Jena a semantic web framework for java, http://jena.sourceforge.net/
23. OASIS: Core and hierarchical role based access control (RBAC) profile of XACML v2.0, http://docs.oasis-open.org/xacml/2.0/access_control-xacml-2.0-rbac-profile1-spec-os.pdf
24. Smith, M.K., Welty, C., McGuinness, D.L.: OWL web ontology language guide (2004), http://www.w3.org/TR/owl-guide/
25. Stanford: Protégé, http://protege.stanford.edu/
26. Bechhofer, S.: The DIG description logic interface: DIG/1.0

# Automatic Rule Generation Based on Genetic Programming for Event Correlation

G. Suarez-Tangil, E. Palomar, J.M. de Fuentes, J. Blasco, and A. Ribagorda

Department of Computer Science – University Carlos III of Madrid
Avda. Universidad 30, 28911 Leganes, Madrid
gtangil@pa.uc3m.es, {epalomar,jfuentes,jbalis,arturo}@inf.uc3m.es

**Abstract.** The widespread adoption of autonomous intrusion detection technology is overwhelming current frameworks for network security management. Modern intrusion detection systems (IDSs) and intelligent agents are the most mentioned in literature and news, although other risks such as broad attacks (e.g. very widely spread in a distributed fashion like botnets), and their consequences on incident response management cannot be overlooked. Event correlation becomes then essential. Basically, security event correlation pulls together detection, prevention and reaction tasks by means of consolidating huge amounts of event data. Providing adaptation to unknown distributed attacks is a major requirement as well as their automatic identification. This positioning paper poses an optimization challenge in the design of such correlation engine and a number of directions for research. We present a novel approach for automatic generation of security event correlation rules based on Genetic Programming which has been already used at sensor level.

**Keywords:** Event Correlation, Rule Generation, Genetic Programming, Network Security Management.

## 1 Introduction

Nowadays, network security management is a critical task which involves different security data sources, e.g. IDSs, firewalls, server logs, to name a few. One the one hand, these sources (known as sensors) produces high amounts of heterogeneous information, generally difficult to understand. Thus, cooperation among these sensors becomes essential for security information management. On the other hand, modern attacks pose a challenge to the network infrastructure as these attacks may be not noticed when inspecting each one separately. Event correlation was therefore conceived as a palliative for both problems.

Security information Management (SIM) systems help to gather, organize and correlate security network information, i.e. reducing the amount of time spent by security administrators. OSSIM [1] is an open source SIM implementation which centralizes the detection and monitoring of the security events within an organization. Nevertheless, correlation techniques included in OSSIM are not able to efficiently detect broad attacks such as Distributed Denial of Service (DDoS).

In this context, we are therefore facing the classical problem of providing optimization to a data mining analysis. Many disciplines exist within data mining; those are particular advantageous when connected to artificial intelligence algorithms. Indeed, Genetic Programming (GP) introduces several interesting properties to deal with building complex event correlation rules: Rules can be represented as computer programs with length variability.

**Our contribution:** In this positioning paper, we elaborate on applying GP to optimize log correlation methods. Specifically, we introduce the main aspects of our current work, which is aimed at building optimized OSSIM rules by means of the application of evolutionary computation.

The rest of the paper is organized as follows. In Section 2 we outline emergent problems related to current IDSs and event correlation techniques. Section 3 describes the foundations of our case study, i.e. automatic generation of correlation rules for the OSSIM. Finally, in Section 4 we establish the main conclusions as well as the immediate future work. As this is a work in progress, each section intents to describe the position of this line of research in each one of the corresponding areas.

## 2 Related Work

Several disciplines are related to security event correlation. Here we describe the most representatives.

### 2.1 Intrusion Detection

As stated in [2], IDSs such as SNORT are not enough to detect complex attacks over a network. Depending on where the IDS is placed they will detect some things and skip some others as well as valuable information logged in other devices such as firewalls, routers, web servers, operating system logs... is also missed.

### 2.2 Event Correlation Techniques

To solve the problem originated by the huge amount of events it is necessary to look into all the implicated devices over the enterprise network. Two possible solutions in this way have been proposed: Data Aggregation (centralization of the information) and Event Correlation (combination of multiple events into one relevant event).

Among its advantages, data aggregation provides a centralized management as well as a unification of the security events into a single format; whereas the high volume of the gathered information can be a disadvantage. On the other hand, we can distinguish the following event correlation mechanisms such as those based on statistics, bayesian inference, alert fusion and data correlation (Micro and Macro correlations).

In fact, the latter shows promise as a security event correlation since other models mentioned before are not especially appropriated [2].

Data correlation has to deal with performance issues in a similar way to data mining classic problems. Instead, the application of artificial intelligence techniques involves amount of properties such as flexibility, adaptability, pattern recognition, efficiency [3], to name a few.

An interesting approach to event correlation at the application layer is proposed in [4] in which involving Network, Kernel and Application Layer an efficient correlation is performed. Additionally, authors in [5] use probabilistic correlation techniques, i.e. Bayesian networks, to increase the sensitivity, reduce false alarms and improve log report quality. The correlation is carried out in both sensor and alert models.

## 2.3 OSSIM Correlation

OSSIM has the ability to correlate attacks using both sequence of events and heuristic algorithms. In the former, the correlation engine [6] generates alerts when a detected event matches the pattern described by a rule. These rules are called "detector rules". Once a detector rule is triggered, OSSIM can start monitoring the sensors providing indicators to the correlation engine: this is done by using "monitoring rules".

Contrarily, correlation engine based on heuristic algorithms generates an indicator or snapshot of the global security state of the network as result of event agregation.

Nevertheless, attack signatures must be defined in advance in both methods. Moreover, classic methods lack of efficiency on pattern recognition. For instance, open problems focus on the application of artificial intelligence to provide Data Mining optimization. Evolutionary algorithms are especially suitable for those problems in which a cost–effort trade-off exists. Event correlation is a good example –achieving the optimal solution is feasible, but it is not affordable at all–. Time and computational savings lead us to employ AI-based techniques.

Finally, more robust event correlation engine is provided by BITACORA [7].

## 2.4 Evolutionary Computation

Evolutionary Computation (EC) [8] is a branch of the Artificial Intelligence paradigm which is based on the species evolution theory. Each iteration of an evolutionary algorithm execution (also known as generation) represents a set of solutions. Afterwards, this set is evaluated, and the best individuals will be selected. The basic genetic operators are used over these individuals to produce next generations. This way, an evolutionary algorithm can be described as a supervised search technique.

Genetic programming [9] is a EC-based methodology, where individuals are represented as computer programs. This fact leads to interesting advantages, i.e. individuals can be defined in a flexible representation. The use of a variant of standard GP which is called Gene Expression Programming (GEP) has been proposed in [10]. In this work, authors state that size of populations and the number of generations are critical, sometimes too high, thus these incur a high cost in term of the search space and time. GEP, Linear GP and Multi-Expression Programming (MEP) are analyzed and used to extract IDS rules in [11]. Nevertheless, to the best of our knowledge, GP has not been applied to event correlation. Indeed, GP is especially suitable on this domain due to the intrinsic characteristics of correlation rules i.e. they have variable lengths and may be expressed by computer routines.

## 3 Applying Genetic Programming to Event Correlation

We now describe how to apply GP [12] to event correlation. In particular, we focus on the automatically generation of correlation rules, once security logs have been centralized in the system intrusion management (SIM) provided by OSSIM framework.

### 3.1 Experimental Environment

Briefly, we now describe the emulation environment. First, we collect SIM alerts that OSSIM Server generates as the result of distributed denial of service (DDoS) attacks. These attacks are launched from a botnet located at the same environment. The collected DDoS traces will be finally used as a training set to evaluate the individuals in each generation. Note that our simulations do not learn in real time. Next sections overview all the stages of our experiments.

The development environment is based on ECJ framework, i.e. a Java-based Evolutionary Computation toolkit [13]. ECJ assists in the deployment of genetic program setting the challenge to the definition of the individual, the fitness function and other parameters such as the number of individuals.

Specifically, we use Standard GP where an individual is represented as a "forest of trees", and the genome (i.e. individual's genetic code) is represented as a tree. Tree's nodes are operator functions whereas tree's terminals are the genome's selected attributes. For instance, from our simulations, results especially rely on three building blocks: (i) the representation of the solution, (ii) how the fitness function represents the objective of the problem and, (iii) how the crossover and mutation operators are defined. We further elaborate on this concerns in the following sections.

### 3.2 Preliminary Format Definition

As the result of the evaluation of the alerts stored on the OSSIM, we have selected the following attributes to be introduced on our Genetic Program. These are the following: Destination IP, Destination Port, Source IP, Source Port, Time stamp and Type of event.

Our goal is to design our Genetic Program as much general as possible. We would like to leave the door open for the future so we can extend our rule generator to other Security Information Manager such as BITACORA.

For this purpose we have decided to give a normalized format to the events. OSSIM output and our GP input are expressed in IDMEF (Detection Message Exchange Format) format [14].

Also, it is necessary to give a normalized type for the event type. The problem is that each IDS follows its own taxonomy. We have solved this problem by parsing the event type string and taking the main keywords that can be found on the event type. Examples on such keywords are "FTP","NMAP", etc.

### 3.3 Representation of the Individual

At first we represented the individuals with the parameters described in 3.2. These parameters come straight forward from the attributes used by some sensor. The use of this attributes provides us the ability of extending the correlations to other systems. A problem to be solved is that we need more attributes to compound a rule; some of the attributes are specific from OSSIM such as the reliability, the occurrence and many others.

As non terminal nodes we have selected the Boolean functions AND and OR. We present two open alternatives for terminal node representation. Terminal nodes are defined as ERC (Ephemeral Random Constants) nodes to guarantee the coherence of the randomly selected and crossed values.

**First Approach: Rule as an ERC:** Let's specify first the resultant work we are looking for. We want our genetic program to automatically generate valid directives for the OSSIM correlator engine.

One directive is compound by many rules in the relation: Directive ⟶ (Rule+).

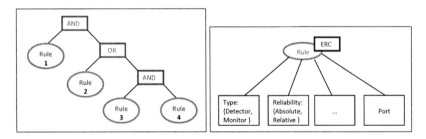

**Fig. 1.** (a) Genetic tree directive (b) A Rule as a ERC terminal element

As all the rules can have the same arguments and we want to generate rules that operate with other rules, (Figure 1 a) the best encoding of the attributes that we can do is to wrap them into an ERC data type. This ERC will define a Rule (Figure 1 b)

A major problem for this representation is that the all the attributes from the rule have to be randomly generated and is much more complicated (in terms of computation) to generate a 'good' rule.

The search space is bigger since we have to find sets of attributes (a rule), with all its possible combinations, that have elements in common with other attributes of another rule.

**Second Approach: Attributes as an ERC:** This second solution tries to solve the problem encountered on the first approach. In this case the terminal elements are the attributes.

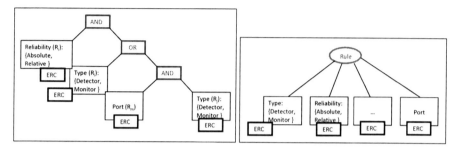

**Fig. 2.** (a) Attributes as an ERC terminal elements. (b)Genetic tree directive

A major problem for this representation is that we are treating the attributes as terminal elements so a possible solution is a tree where attributes of a rule N can operate with attributes of the rule M, where N can be equal to M; this is not always a valid OSSIM directive.

### 3.4 Genetic Operators

The following genetic operators are performed: *reproduction, mutation and crossover*. The *reproduction* operator is responsible of the clonation of the individuals: These cloned individuals are randomly modified in the *mutation* operator. During *mutation* and *crossover* operators we have use default ECJ framework values.

ECJ provides a simple way to implement a specific crossover and mutation operators. Depending on the accuracy of the results, this might be implemented.

### 3.5 Training and Fitness Function

Each round of the algorithm will be evaluated with a training set logged in our training environment which contains events from specific attacks.

Additionally, the fitness function will be based on the risk assessment as defined by OSSIM.

Therefore, detector rules follow a logical evaluation of the attributes. Each time an event is detected by a rule, its parameter *Reliability* is incremented. This change also affects the Risk calculation. In this situation, considering the Risk parameter for building the fitness function seems to be suitable. Since the monitoring rules (Eq. 1) follow a heuristic algorithm which is defined as the risk that the monitorized asset might be suffering for a specific thread (reliability of the possible attack) and the priority of the directive.

$$Risk = \frac{(Asset \cdot Priority \cdot Reliability)}{25} \quad (1)$$

## 4  Conclusions and Research Directions

In this positioning paper, we have studied the suitability of evolutionary computation techniques to improve the efficiency of current security event correlation mechanisms. In fact, genetic programming lets us manage the problem in a very natural way: correlation rules can be built based on the evolution of individuals as a mechanism to discover event "proximity". We have outlined our findings in building OSSIM correlation directives by means of GP. Nevertheless, our approach faces up some research problems, such as an appropriate population representation and the generation of a suitable training set.

In the short term, we further elaborate on the task of correlation and pattern matching which is still one for a strong analysis concern. Finally, as future work, we are evaluating the viability of the presented representation of the problem, which seems to be a completely original approach to tackle the security event correlation problem, by means of system emulation.

Our hope is that this paper will, directly or indirectly, inspire new directions in efficiently automatizing security event correlation.

## Acknowledgements

This work is supported by CDTI, Ministerio de Industria, Turismo y Comercio of Spain in collaboration with Telefonica I+D, project SEGUR@ with reference CENIT-2007 2004.

## References

1. OSSIM: Open source security information management (2009), http://www.ossim.net/whatis.php
2. Center for Education and Research in information Assurance and Security of Purde University: CERIAS Security Seminar Archive - Intrusion Detection Event Correlation: Approaches, Benefits and Pitfalls, Center for Education and Research in information Assurance and Security of Purde University (March 2007)
3. Tjhai, G.: Intrusion detection system: Facts, challenges and futures (March 2007), http://www.bcssouthwest.org.uk/presentations/GinaTjhai2007.pdf
4. Rice, G., Daniels, T.: A hierarchical approach for detecting system intrusions through event correlation. In: IASTED International Conference on Communication, Network, and Information Security, Phoenix, USA (November 2005)
5. Valdes, A., Skinner, K.: Probabilistic alert correlation. In: Proceedings of the 4th International Symposium on Recent Advances in Intrusion Detection, pp. 54–68 (2001)
6. Karg, D.: OSSIM Correlation engine explained (2004), http://www.ossim.net/docs/correlation_engine_explained_rpc_dcom_example.pdf
7. Bitacora: System of centralization, management and exploitation of a company's events, http://www.s21sec.com/productos.aspx?sec=34
8. Fogel, L.J., Owens, A.J., Walsh, M.J.: Artificial Intelligence through Simulated Evolution. John Wiley, New York (1966)

9. Koza, J., Poli, R.: Introductory Tutorials in Optimization and Decision Support Techniques. Springer, UK (2005)
10. Tang, W., Cao, Y., Yang, X., So, W.: Study on adaptive intrusion detection engine based on gene expression programming rules. In: CSSE International Conference on Computer Science and Software Engineering, Wuhan, China (December 2008)
11. Eskin, E., Arnold, A., Prerau, M., Portnoy, L., Stolfo, S.: A geometric framework for unsupervised anomaly detection: Detecting intrusions in unlabeled data. In: Applications of Data Mining in Computer Security. Kluwer, Dordrecht (2002)
12. Mukkamala, S., Sung, A.H., Abraham, A.: Modeling intrusion detection systems using linear genetic programming approach. In: Orchard, B., Yang, C., Ali, M. (eds.) IEA/AIE 2004. LNCS (LNAI), vol. 3029, pp. 633–642. Springer, Heidelberg (2004)
13. Luke, S., Panait, L., Balan, G., Paus, S., Skolicki, Z., Popovici, E., Sullivan, K., Harrison, J., Bassett, J., Hubley, R., Chircop, A.: A java-based evolutionary computation research system, http://cs.gmu.edu/~eclab/projects/ecj/
14. Debar, H., Curry, D., Feinstein, B.: Ietf rfc 4765 - the intrusion detection message exchange format (March 2007), www.ietf.org/rfc/rfc4765.txt

# Learning Program Behavior for Run-Time Software Assurance

Hira Agrawal[1], Clifford Behrens[1], Balakrishnan Dasarathy[1], and Leslie Lee Fook[2]

[1] One Telcordia Drive, Piscataway, NJ 08854, USA
{hira,cliff,das}@research.telcordia.com
[2] 4 RobinsonVille, Belmont, Port of Spain, Trinidad, WI
leslie@smbtrinidad.com

**Abstract.** In this paper we present techniques for machine learning of program behavior by observing application level events to support runtime anomaly detection. We exploit two key relationships among event sequences: their edit distance proximity and state information embedded in event data. We integrate two techniques that employ these relationships to reduce both false positives and false negatives. Our techniques consider event sequences in their entirety, and thus better leverage correlations among events over longer time periods than most other techniques that use small, fixed length sliding windows over such sequences. We employ cluster signatures that minimize adverse effects of noise in anomaly detection, thereby further reducing false positives. We leverage state information in event data to summarize loop structures in sequences which, in turn, leads to better classification of program behavior. We have performed initial validations of these techniques using Asterisk®, a widely deployed, open source digital PBX.

**Keywords:** program behavior learning, unsupervised learning, anomaly detection, intrusion detection, automata generation.

## 1 Introduction

Defects persist in today's software despite extensive testing and verification. These defects can cause security vulnerabilities, failures, downtime, and performance problems. Moreover, many large applications consist of components from different vendors, whose features may interact in unexpected ways and cause undesirable behavior. These problems are further exacerbated in distributed and service-oriented environments where software components can be sourced dynamically at runtime. Clearly, comprehensive testing and analysis of complex, highly configurable systems—prior to their deployment—isn't always feasible due to the combinatorial complexity of feature interactions, irreproducibility of certain bugs, or the cost and delays associated with further testing. Consequently, runtime system assurance solutions are needed as a last line of defense against software defects.

Our research in this area has primarily concentrated on integration of specification and learning-based approaches for identifying potentially disabling system behavior.

Since manual specification of desired program behavior in a machine-understandable form is practically impossible for large, highly configurable systems, our approach employs machine learning of program behavior. The general problem of learning program behavior from observed events, however, is computationally hard, and practical solutions often suffer from too many false positives and false negatives. Our goal is to produce an anomaly detection solution that attempts to minimize both false positives and false negatives. To achieve this, our approach couples two complementary techniques—event sequence clustering and automata generation—that leverage two types of relationships in observed events: event sequence proximity and system state. Another key aspect of our approach is that it performs anomaly detection using application level events, and therefore, leverages rich application semantics associated with such events. This is in contrast with most other techniques that focus on operating system level events such as system calls. Another advantage of our approach is that it can be used to address multiple software dependability traits, such as performance, reliability, availability, and security—including attacks that exploit program vulnerabilities—under a single framework, as the root causes behind failures associated with all these attributes are hard to discover defects and unanticipated interactions among features and components of the target system.

We describe the two key techniques underlying our approach—event sequence clustering and automata generation—in Sections 2 and 3, respectively. We discuss our proposed hybrid approach combining these two techniques in Section 4. Related work is discussed in Section 5, and concluding remarks are provided in Section 6.

## 2 Edit Distance Based Clustering

The core of our approach is an unsupervised learning technique, which uses statistical clustering of event sequence strings based on edit distances among those strings. The main intuition here is that program inputs that undergo similar processing by the system should produce similar event sequences. To test and refine our approach, we have been experimenting with data generated by Asterisk® [1], an open source PBX with hundreds of thousands of deployments worldwide. While we will explain our approach in the context of Asterisk, it is by no means limited to anomaly detection for that application. Fig. 1 and Fig. 2 illustrate the core steps in our approach.

First, we compute edit distances among all event sequences observed during the training phase, each modeled as a character string composed from a customized alphabet. Characters in this alphabet are mapped to call events in Asterisk. We use Levenshtein distance which computes the distance between two event sequence strings as the minimum number of simple edit operations, viz., character insertions, deletions, and substitutions, that are required to convert one event string to another. Next we reduce the dimensionality of the derived distance matrix with principal components analysis (PCA). PCA extracts components of the given data that are orthogonal to one another and orders them so that each successive component accounts for as much of the remaining variability in the data as possible.

Then a SCREE plot showing the amount of variability in the distance matrix accounted for by each principal component is derived, as illustrated in Fig. 1. This plot indicates that most of the variability in the distance matrix can be accounted for

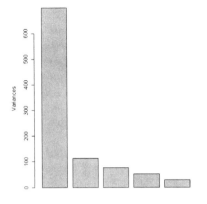
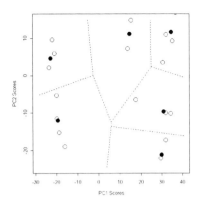

**Fig. 1.** Principal components analysis based application event sequence clustering

**Fig. 2.** Clustering of component scores

by the first two principal components, as there exists an "elbow" in the plot after these two components have been extracted. Fig. 2 shows a bivariate plot of event sequences scored on these two principal components. Although this test set contained one thousand calls, it produced only 19 unique call sequences. Many calls resulted in the same event sequence, because they differed only in factors such as the time the call was made, the duration of the call, and the calling and called parties involved. We then apply the K-means clustering algorithm to the first two component scores to discover clusters of related call types based on their spatial proximity in this two dimensional space, also shown in Fig. 2. Hollow circles in this figure represent call event sequences and filled circled represent the corresponding cluster centroids.

We then derive a signature for each cluster, computed as the longest common subsequence (LCS) [5] shared by all event sequences in that cluster. LCS differs from the longest common substring in that characters included in the former need not occur consecutively in the event sequence; they just need to occur in the same relative order. The signature derived from LCS offers the advantage of ignoring noise in event sequences during anomaly detection, as events in a signature do not need to appear together in the target event sequence.

We believe that clustering based on edit distance provides a solid basis for anomaly detection. Clusters we computed from the above data indicate that they all have application-level meanings. For instance, the cluster in the top left hand corner of Fig. 2, contains all calls that were answered and completed normally. The cluster in the top right hand corner, on the other hand, includes all calls that were sent to voicemail. Clustering performed on call sequences obtained from a larger, production Asterisk installation also confirms these encouraging results.

Note that the number of clusters involved may vary from one installation of the system to another. This is because the manner in which the system is configured and used may vary from one place to another. In the case of Asterisk, each installation will have its own, unique "dial plan." An Asterisk dial plan is essentially a script which dictates how various calls, such as incoming and outgoing calls, are handled at that particular installation. Moreover, a domain expert may be required to determine the appropriate number (K) of clusters for each installation.

Once signatures have been derived for all clusters, we also derive probability distributions of edit distances of all sequences from their cluster signatures, which, in turn, are used to set thresholds for anomaly detection. If, for example, 95% of sequences in a cluster are within a distance, say, $d$, from their cluster signature, then $d$ may be set as a threshold for anomaly detection for that cluster. In this case, all new sequences placed in that cluster that are farther than distance $d$ from its signature will be classified as anomalous. Note that different clusters may have different distributions, and hence may be assigned different thresholds for detecting anomalous sequences assigned to those clusters.

We tested this anomaly detection technique on the data underlying Fig. 2, by performing clustering on several new call event sequences. One of these sequences was flagged as anomalous, whereas others were not. Upon closure examination of these sequences, we found that the anomalous call was, indeed, a totally new type of "call" that had not been seen in the training data: one where the SIP address of the called party referred to a nonexistent SIP proxy server! Other calls were not flagged as anomalous, as they were adequately represented in the training set, and their distances from the corresponding cluster centroids were within their respective thresholds.

## 3 Leveraging State Information

Clustering of event sequences based on edit distances between their string representations is very sensitive to the lengths of those sequences, in the sense that edit distance between two sequences of widely different lengths will be large, and hence those sequences will likely belong to different clusters. This poses a problem for sequences containing subsequences that represent loop iterations. Edit distance between two sequences that are otherwise identical, except for number of iterations of a shared loop, will be large if their iteration counts differ significantly. Consider, for example, multi-party conference calls. Their call sequences will contain multiple subsequences representing callers joining and leaving those calls. Ideally, such sequences should belong to the same cluster, even though they differ in number of call participants.

A loop body may also contain internal branching structure, and different iterations of the loop may follow paths containing different branches. Detection of recurring event substrings when there is no internal branching involved can be handled easily using techniques such as suffix tries [12]. In such cases, each substring occurrence may be replaced by a corresponding "macro" character. This approach does not work when loop bodies contain multiple internal paths, as is often the case. These sequences can, however, be summarized using regular expressions. Such loop structures arise frequently in call processing sequences. A common example where complex loop structures arise in call processing is interactive voice response menus. We, therefore, need a mechanism to learn regular expression patterns from call event sequence strings. The problem of learning a regular expression from a given set of event sequences is the same as that of inferring a finite state machine (FSM) that accepts those sequences. Hence, we need an algorithm—an FSM synthesizer—that generates a finite state machine for the given sequences, using as few states as is feasible. It is, however, not feasible to infer a compact FSM when only the set of acceptable strings is provided. This is because, without also knowing the set of unacceptable strings, the algorithm has no way of knowing when it is over-generalizing. In other words, an FSM synthesizer will produce too many false negatives if it is only shown acceptable strings. So, we need both positive (acceptable) and negative (unacceptable) samples. In anomaly detection, however, we only have positive samples. Moreover, finding a deterministic finite-state automaton (DFA) with minimum number of states, given both sets of acceptable and unacceptable strings, is an NP-hard problem [4].

In view of this, we take a practical approach based on deriving state information from data that is contained in the events themselves, i.e., we discover and exploit clues about "internal states" of an application from the events at the time they were emitted. In case of Asterisk, we were able to obtain state information from many Asterisk events to generate finite state machines from observed sequences of call events. Fig. 3 shows a fragment of a finite state machine generated using this approach. It represents segments of call sequences related to an interactive voice response (IVR) menu, such as those that are typically encountered when retrieving one's voice mail.

Some of the Asterisk events contain a field that reports the status of the telephone line involved, using terms like "Ringing", "Busy", "Up", and "Down". This field provides a direct clue about the internal state of the corresponding phone line. Such states, however, are shared by phone lines involved in all types of calls: incoming calls, outgoing calls, internal calls, conference calls, IVR sessions, etc. Therefore we combine this field with some other fields such as "context" and "extension" to infer a set of application "states". Even when state clues like these are lacking in the observed event sequences, it may be possible to exploit "white box" clues. These might include line numbers of program statements that produced those events or the value of the program counter when those events were emitted.

## 4 A Hybrid Approach

If FSMs like the one discussed above can be generated from event sequences, do we still need clustering based anomaly detection? The answer is yes, because, in practice, FSMs generated in this way will still give rise to many false negatives. This is because the state information we can obtain from event data is far from being complete. For example, the line number of a statement that emits an event does not capture the complete state of the program at that time. To capture its true state, one would need to capture values for all variables in the program at that instant—an infeasible task given the combinatorial explosion it would cause in the number of possible states. For this reason, we don't use the generated FSM as the sole basis for anomaly detection. Instead, we use a hybrid approach that combines event sequence clustering with FSM generation. Our approach uses the generated FSM to collapse all loops in the observed event sequences into their corresponding, succinct regular expression based representation before those sequences are clustered.

Our approach collapses all loop structures in event sequences into equivalent summary strings using the learned FSM. As an example of this linearization process,

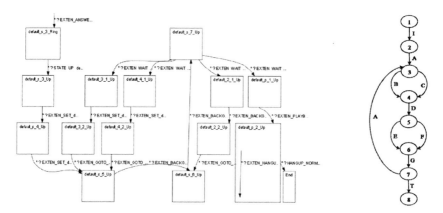

**Fig. 3.** An example of finite state machine synthesis

**Fig. 4.** Another example

assume that the simple FSM shown in Fig. 4 was generated from observed event sequences. The letters next to arrows in this figure represent events and the numbered circles represent the learned states. The following string represents the regular expression that corresponds to event sequences accepted by this FSM: I(A(B|C)D(E|F)G)+T.

In linearization, our goal is to preserve the original sequence to the degree possible. If a call event sequence contains no loops, as in the sequence IABDEGT, we leave it unmodified. If the original sequence were IABDEGABDEGT, we would replace it with I(ABDEG)2T to capture the fact that the subsequence ABDEG is repeated twice. Notice that the edit distance between I(ABDEG)2T and IABDEGT is much less than that between IABDEGABDEGT and IABDEGT. Thus, I(ABDEG)2T has a much better chance of being clustered with IABDEGT. If, however, the original sequence were IABDEGACDEGT, we would replace it with I(A(B|C)DEG)2T to show the use of alternate paths following the event A, as well as the number of loop iterations. Notice that we are not using E|F in this regular expression representation, as the event D was followed by event E during both iterations. As you can see, this linearization takes into account arbitrary looping but preserves proximity to the extent possible among various event strings for clustering purposes.

## 5 Related Work

A lot of work has been performed on signature or pattern-based misuse detection, where application event streams are searched against known attack patterns (see [9] for one such work). Such techniques, however, are not appropriate at detecting new, previously unknown attacks. Some work has also been done where, instead of modeling invalid program behavior, its desired behavior is specified [7], and any deviations from the latter are flagged as possible anomalies or attacks. Requiring such specifications to be provided for large, complex systems, however, is not practical.

Most work in host-based anomaly or intrusion detection has focused on operating system level events such as system calls [6, 11], and it is mostly based on the "n-gram" or the sliding window technique. As these methods learn allowable patterns in sequences of small, fixed lengths, they fail to take into account correlations among events over longer time periods.

Clustering is an extensively researched topic in anomaly detection in networks (see [8] for a survey paper on the topic). It is typically applied to feature spaces consisting of various discrete valued attributes for detecting outliers. They are not applied to event sequences. One area where sequence similarity measures have been extensively used is bioinformatics, specifically in the analysis of newly sequenced genes [10].

Our anomaly detection technique also depends on learning the automaton behind the observed behavior. This problem has been well studied for over forty years, both on its intractability aspects [4] and on providing acceptable solutions in specific domains, e.g., software process modeling [3]. Wee et al. use FSM generation to model relationships among operating system level events and employ edit distances among event sequences [12]. They do not, however, use clustering or deal with arbitrary recursion in strings. Bhatkar et al. [2] propose FSM generation using values of program counters when system-call events are emitted. This approach is not well suited for dynamically-linked programs, and it will, if used alone, lead to many false negatives, as we observed earlier.

## 6 Conclusions

We have presented a new technique for learning program behavior for anomaly detection that leverages two types of relationships in observed event sequences— event proximity and system state. Furthermore, it integrates them in a unique way that achieves both low false positives and low false negatives. We have validated this technique using Asterisk, an open source PBX with thousands of deployments worldwide. We were also able to demonstrate that learning state information from events is a practical and computationally feasible approach for inferring automata from the event strings. We used finite state machine inference to recognize complex iterations—with internal branching structures—in event strings, so they can be replaced with their regular expression based summary strings. This enables proximity among event sequences that are similar in functionality but vary only in the number of loop iterations to be preserved.

In future, we plan to extend this approach so it overcomes some of its current limitations. The anomaly detection approach discussed in this paper requires that we wait for a complete event sequence to be known before we can determine whether or not it represents an anomaly. But such "after-the-fact" determination of an anomaly limits its usefulness in terms of actions one can take to prevent its harmful impact on the system. We would, ideally, like to know if the event sequence in progress likely represents an anomaly before we have reached the end of that sequence, so we can take an appropriate preventive action. A simple extension we are considering for detecting an anomaly—without having to wait until the end of a sequence—is to treat all prefixes, longer than a certain minimum threshold, of all sequences in the original training set as independent entries in the new training set, and perform edit distance based clustering on this new set.

Our anomaly detection approach only looks at positions of events relative to one another within their enclosing sequences. But often times, the time elapsed between consecutive events within a sequence may be just as important for determining if that sequence represents an anomaly. In other words, two sequences may have identical event strings, yet one of them may be anomalous if, for example, the time durations between certain events are unusually large compared to what was observed during the learning phase. Therefore, in addition to edit distance proximity and state information, we are also exploring use of time duration information between consecutive events in anomaly detection. We plan to investigate event-duration algorithms such as Random Observation Time Hidden Markov Models. Such algorithms have already proven successful for classifying event sequence data in other applications such as speech recognition, behavior tracking, and health surveillance.

## References

1. Asterisk open source digital PBX, http://www.asterisk.org
2. Bhatkar, S., Chaturvedi, A., Sekar, R.: Dataflow Anomaly Detection. In: Proceedings of the IEEE Symposium on Security and Privacy, pp. 48–62 (2006)
3. Cook, J., Wolf, A.L.: Discovering Models of Software Processes from Event-Based Data. ACM Trans. Software Engineering and Methodology 7(3), 215–249 (1998)
4. Gold, E.: Language identification in the limit. Inf. Control 10, 447–474 (1967)
5. Gusfield, D.: Algorithms on Strings, Trees, and Sequences: Computer Science and Computational Biology. Cambridge University Press, Cambridge (1997)
6. Hofmeyr, S.A., Somayaji, A., Forrest, S.: Intrusion Detection using Sequences of System Calls. Journal of Computer Security 6, 151–180 (1998)
7. Ko, C.: Logic induction of valid behavior specifications for intrusion detection. In: Proc. IEEE Symposium on Security and Privacy (2000)
8. Lazarevic, A., Ertoz, L., Ozgur, A., Srivastava, J., Kumar, V.: A comparative study of anomaly detection schemes in network intrusion detection. In: SDM (2003)
9. Lindqvist, U., Porras, P.A.: eXpert-BSM: A Host-based Intrusion Detection Solution for Sun Solaris. In: Proc.17th Annual Computer Security Applications Conference, pp. 240–251 (2001)
10. Needleman, S.B., Wunsch, C.D.: A general method applicable to the search for similarities in the amino acid sequence of two proteins. Journal of Molecular Biology 48, 443–453 (1970)
11. Tandon, G., Chan, P.: Learning Rules from System Call Arguments and Sequences for Anomaly Detection. In: Workshop on Data Mining for Computer Security, pp. 20–29 (2003)
12. Wee, K., Moon, B.: Automatic generation of finite state automata for detecting intrusions using system call sequences. In: Gorodetsky, V., Popyack, L.J., Skormin, V.A. (eds.) MMM-ACNS 2003. LNCS, vol. 2776, pp. 206–216. Springer, Heidelberg (2003)

# Multiagent Systems for Network Intrusion Detection: A Review

Álvaro Herrero and Emilio Corchado

Department of Civil Engineering, University of Burgos
C/ Francisco de Vitoria s/n, 09006 Burgos, Spain
{ahcosio,escorchado}@ubu.es

**Abstract.** More and more, Intrusion Detection Systems (IDSs) are seen as an important component in comprehensive security solutions. Thus, IDSs are common elements in modern infrastructures to enforce network policies. So far, plenty of techniques have been applied for the detection of intrusions, which has been reported in many surveys. This work focuses the development of network-based IDSs from an architectural point of view, in which multiagent systems are applied for the development of IDSs, presenting an up-to-date revision of the state of the art.

**Keywords:** Multiagent Systems, Distributed Artificial Intelligence, Computer Network Security, Intrusion Detection.

## 1 Introduction

Firewalls are the most widely used tools for securing networks, but Intrusion Detection Systems (IDSs) are becoming more and more popular [1]. IDSs monitor the activity of the network with the purpose of identifying intrusive events and can take actions to abort these risky events. A wide range of techniques have been used to build IDSs, most of the reported and described in previous surveys [2], [3], [4], [5], [6], [7], [8], [9], [10], [11], [12], [13], [14], [15].

On a more general context, the actual demands of effectiveness and complexity have caused the development of new computing paradigms. Agents and multiagent systems (MAS) [16] are one of these new paradigms. The concept of agent was originally conceived in the field of Artificial Intelligence (AI), evolving subsequently as a computational entity in the software engineering field. From the software standpoint, it is seen as an evolution to overcome the limitations inherent to the object oriented methodologies. Up to now, there is not a strict definition of agent [17]. In a general AI context, a rational agent was defined [18] as anything that perceives its environment through sensors and acts upon that environment through effectors. In a more specific way, a software agent has been defined as a system with capacity of adaptation and provided with mechanisms allowing it to decide what to do (according to their objectives) [19]. Additionally, from a distributed AI standpoint, it was defined [20] as a physical or virtual entity with some features: capable of acting in an environment, able of communicating directly with other agents, possessing resources of its own, and some others.

It is in a multiagent system (MAS) that contains an environment, objects and agents (the agents being the only ones to act), relations between all the entities, a set of operations that can be performed by the entities and the changes of the universe in time and due to these actions" [20]. From the standpoint of distributed problem solving [21] a MAS can be defined as a loosely coupled network of problem solvers that work together to solve problems that are beyond the individual capabilities or knowledge of each problem solver. According to [22], the characteristics of MASs are:

- Each agent has incomplete information, or capabilities for solving the problem, thus each agent has a limited viewpoint.
- There is no global system control.
- Data is decentralized.
- Computation is asynchronous.

As a consequence of that, agents in a MAS are driven by their own objectives as there is not a global control unit. They take the initiative according to their objectives and dynamically decide what to do or what tasks other agents must do.

Agents and multiagent systems have been widely used in last years, not always being the most appropriate solution. According to [23], there is a number of features of a problem which point to the appropriateness of an agent-based solution:

- The environment is open, or at least highly dynamic, uncertain, or complex.
- Agents are a natural metaphor. Many environments are naturally modelled as societies of agents, either cooperating with each other to solve complex problems, or else competing with one-another.
- Distribution of data, control or expertise. It means that a centralised solution is at best extremely difficult or at worst impossible.
- Legacy systems. That is, software technologically obsolete but functionally essential to an organisation. Such software cannot generally be discarded (because of the short-term cost of rewriting) and it is often required to interact with other software components. One solution to this problem is to wrap the legacy components, providing them with an "agent layer" functionality.

Since its inception in the 1980s, IDSs have evolved from monolithic batch-oriented systems to distributed real-time networks of components [10]. As a result, new paradigms have been designed to support such tools. Agents and multiagent systems are one of the paradigms that best fit this setting as ID in distributed networks is a problem that matches the above requirements for an agent-based solution. Furthermore, some other AI techniques can me combined with this paradigm to make more intelligent agents.

This paper surveys and chronologically analyses previous work on multiagent systems for network intrusion detection (Section 2), emphasizing the mobile agent approach (Section 3).

## 2 IDSs Based on Agents

One of the initial studies under this frame was JAM (Java Agents for Metalerning) [24]. This work combines intelligent agents and data mining techniques. When applied to the ID problem, an association-rules algorithm determines the relationships between the different fields in audit trails, while a meta-learning classifier learns the signatures of attacks. Features of these two data mining techniques are extracted and used to compute models of intrusion behaviour.

In the 90's DARPA defined the Common Intrusion Detection Framework (CIDF) as a general framework for IDS development. The Open Infrastructure [25] comprises a general infrastructure for agent based ID that is CIDF compliant. This infrastructure defines a layered agent hierarchy, consisting of the following agent types: Decision-Response agents (responsible for responding to intrusions), Reconnaissance agents (gather information), Analysis agents (analyse the gathered information), Directory/Key Management agents, and Storage agents. The two later provide support functions to the other agents.

AAFID (Autonomous Agents For Intrusion Detection) [26] is a distributed IDS architecture employing autonomous agents, being those defined as "*software agents that perform a certain security monitoring function at a host*". This architecture defines the following main components:

- **Agents:** monitor certain aspects of hosts and report them to the appropriate transceiver.
- **Filters:** intended to be the data selection and abstraction layer for agents.
- **Transceivers:** external communications interfaces of hosts.
- **Monitors:** the highest-level entities that control entities that are running in several different hosts.
- **User interface:** interact with a monitor to request information and to provide instructions.

This architecture does not detail the inner structure or mechanisms of the proposed agents, that use filters to obtain data in a system-independent manner. That is, agents do not depend on the operating system of the hosts. Additionally, AAFID agents do not have the authority to directly generate an alarm and do not communicate directly with each other.

In [27], a general MAS framework for ID is also proposed. Authors suggest the development of four main modules, namely the sniffing module (to be implemented as a simple reflex agent), the analysis module (to be implemented as several agents that keeps track of the environment to look at past packets), the decision module (to be implemented as goal-based agents to make the appropriate decisions), and the reporting module (to be implemented as two simple reflex agents: logging and alert generator agents). These components are developed as agents:

- The sniffing agent sends the previously stored data to the analysis agents when the latter request new data. One analyser agent is created for each one of the attacks to be identified. They analyse the traffic reported from the sniffing module, searching for signatures of attacks and consequently building a list of suspicious packets.

- Decision agents are attack dependant. They calculate the severity of the attack they are in charge from the list of suspicious packets built by analyser agents. Decision agents also take the necessary action according to the level of severity.
- Finally, the logging agent keeps track of the logging file, accounting for the list of suspect packets generated from the decision agents. On the other hand, the alert generator agent sends alerts to the system administrator according to the list of decisions.

Some topics about ID based on MASs are briefly discussed in [28], where a general framework for ID is proposed. Such framework includes the following classes of agents: learning data management agents, classifier testing agents, meta-data forming agents, and learning agents.

SPIDeR-MAN (Synergistic and Perceptual Intrusion Detection with Reinforcement in a Multi-Agent Neural Network) is proposed in [29]. Each agent uses a SOM and ordinary rule-based classifiers to detect intrusive activities. A blackboard mechanism is used for the aggregation of results generated from such agents (i.e. a group decision). Reinforcement learning is carried out with the reinforcement signal that is generated within the blackboard and distributed over all agents which are involved in the group decision making.

An heterogeneous alert correlation approach to ID by means of a MAS is proposed in [30]. In this study alert correlation refers to the management of alerts generated by a set of classifiers, each of them trained for detecting attacks of a particular class (DoS, Probe, U2R, etc.). Although it is a Host-based IDS (HIDS), the main idea underlying the design of this MAS could be also applied to Network-based IDSs (NIDSs). According to the adopted Gaia methodology, roles and protocols are specified in this study. The roles are mapped into the following agent classes:

- **NetLevelAgent (DataSensor role):** in charge of raw data preprocessing and extracting both events and secondary features.
- **BaseClassifiers (DecisionProvider role):** performs source based classification and produces decisions after receiving events from sources. Several subclasses are defined to cover the different predefined types of attacks and the different data sources.
- **Metaclassifiers (DecisionReceiver and DecisionProvider roles):** one agent of this class is instantiated for each one of the attack types. They combine decisions produced by the BaseClassifiers agents of the assigned attack type.
- **SystemMontor (ObjectMonitor role):** visualises the information about security status.

CIDS (Cougaar-based IDS) [31] provides a hierarchical security agent framework, where security nodes are defined as consisting of four different agents (manager agent, monitor agent, decision agent, and action agent) developed over the Cougaar framework [32]. It uses intelligent decision support modules to detect some anomalies and intrusions from user to packet level. The output of CIDS (generated by the Action Agent) consists on the environment status report (IDMEF format [33]) as well as recommendations of actions to be taken against the ongoing intrusive activities. The system employs a knowledgebase of known attacks and a fuzzy inference engine to classify network activities as legitimate or malicious.

PAID (Probabilistic Agent-Based IDS) [34] is a cooperative agent architecture where autonomous agents perform specific ID tasks (e.g., identifying IP-spoofing attacks). It uses three types of agents:

- **System monitoring agents:** responsible for collecting, transforming, and distributing intrusion specific data upon request and evoking information collecting procedures
- **Intrusion-monitoring agents:** encapsulate a Bayesian Network and performs belief update using both facts (observed values) and beliefs (derived values). They generate probability distributions (beliefs) over intrusion variables that may be shared with other agents, which constitutes the main novelty of PAID. Methods for modelling errors and resolving conflicts among beliefs are also defined.
- **Registry agents:** coordinate system-monitoring and intrusion-monitoring agents.

A multiagent IDS framework for decentralised intrusion prevention and detection is proposed in [35]. The MAS structure is tree-hierarchical and consists of the following agents:

- **Monitor agents:** capture traffic, preprocess it (reducing irrelevant and noisy data), and extract the latent independent features by applying feature selection methods.
- **Decision agents:** perform unsupervised anomaly learning and classification. To do so, an ant colony clustering model is deployed in these agents. When attacks are detected, they send simple notification messages to corresponding action and coordination agents.
- **Action agents:** perform passive or reactive responses to different attacks.
- **Coordination agents:** aggregate and analyse high-level detection results to enhance the predictability and efficiency.
- **User Interface agents:** interact with the users and interpret the intrusion information and alarms.
- **Registration agents:** allocate and look up all the other agents.

A MAS comprising intelligent agents is proposed in [36] for detecting probes. These intelligent agents were encapsulated with different AI paradigms: support vector machines, multi-variate adaptive regression, and linear genetic programming. Thanks to this agent-based approach, specific agents can be designed and implemented in a distributed fashion taking into account prior knowledge of the device and user profiles of the network.

By adding new agents, this system can be easily adapted to an increased problem size. Due to the interaction of different agents, failure of one agent may not degrade the overall detection performance of the network.

MOVIH-IDS (Mobile-Visualization Hybrid IDS) [37] is built by means of a MAS that incorporates an artificial neural network for the visualisation of network traffic. It includes deliberative agents characterized by the use of an unsupervised connectionist model to identify intrusions in computer networks. These deliberative agents are defined as CBR-BDI agents [38], [39] using the Case-based Reasoning paradigm [40] as a reasoning mechanism, which allows them to learn from initial knowledge, to interact autonomously with the environment, users and other agents within the system, and to have a large capacity for adaptation to the needs of their surroundings.

## 3 Mobile Agents

Apart from the above works, some others have focused on the mobile-agent approach. That is, agents travelling along different hosts in the network to be monitored. Some issues about the application of mobile agents to ID are further discussed in [41], and examples following this approach are described in this section.

IDA (ID Agent) system [42] is aimed at detecting many intrusions efficiently rather than accurately detecting all intrusions. To do so, it approaches ID from a novel standpoint: instead of continuously monitoring the activity of users, it watches events that may relate to intrusions (MLSI – Mark Left by Suspected Intruders). When an MLSI is detected, IDA collects further information, analyses it and decides whether an intrusion has taken place. To do so, two kinds of mobile agents contribute to the information collection stage: a tracing agent is sent to the host where suspicious activity comes from and once there, it activates an information-gathering agent. Several information-gathering agents may be activated by several different tracing agents on the same target system.

Micael [43] was proposed as an IDS architecture based on mobile agents. Its main difference to previous proposals is the task division. ID tasks are distributed to the following agent kinds: head quarter (centralizes the system's control functions), sentinels (collect relevant information, and inform the head quarter agents about eventual anomalies), detachments (implement the counter-measures of the IDS), auditors (check the integrity of the active agents), and special agents with different duties. By moving throughout the network, the mobile auditor agents can audit each of the defended hosts sequentially.

Mobile agents are applied in [44] to make critical IDS components resistant to flooding DoS and penetration attacks. To do so, the attacked IDS component will be automatically relocated to a different (still operational) host. This relocation is invisible to the attacker who then cannot persist in the attack. Every critical agent has one or more backup agents (maintaining full or partial state information of the agent they are backing up) that reside on distinct hosts within the same domain. When the machine hosting a critical agent is down (whatever the reason is), its backup agents contact each other to decide on a successor that will resume the functions of the original agent. One of the main drawbacks of this solution is that temporarily the network may be partially unprotected while the IDS critical components are moving from one host to another.

SPARTA (Security Policy Adaptation Reinforced Through Agents) is proposed in [45] as an architecture to collect and relate distributed ID data using mobile agents. According to the authors, SPARTA mobile agents enable the distributed analysis, improve the scalability, and increase the fault tolerance. Some security issues about these mobile agents are considered. The required information (interesting host events) for event correlation is locally collected and stored, which is considered a distributed database with horizontal fragmentation. Mobile agents are in charge of collecting the distributed information (matching a given pattern) to answer user queries.

SANTA (Security Agents for Network Traffic Analysis) [46] is proposed as a distributed architecture for network security using packet, process, system, and user information. It attempts to emulate mechanisms of the natural immune system using IBM's Aglets agents. The proposed monitoring agents roam around the machines (hosts or routers) and monitor the situation in the network (i.e., look for changes such as malfunctions, faults, abnormalities, misuse, deviations, intrusions, etc.). These immunity-based agents can mutually recognize each other's activities and implement Adaptive Resonance Theory neural networks and a fuzzy controller for ID. According to the underlying security policies and the information from the monitoring agents, decision/action agents make decisions as to whether an action should be taken by killer agents. These killer agents terminate processes that are responsible for intrusive behaviour on the network.

A distributed ID architecture, completed with a data warehouse and mobile and stationary agents is proposed in [47]. The MAS is combined with a rule generation algorithm, genetic algorithms, and datawarehouse techniques to facilitate building, monitoring, and analysing global, spatio-temporal views of intrusions on large distributed systems. System calls executed by privileged processes are classified after being represented as feature vectors. To do so, different agents are defined:

- **Data cleaner agents:** these stationary agents process data obtained from log files, network protocol monitors, and system activity monitors into homogeneous formats.
- **Low-level agents:** these mobile agents form the first level of ID. They travel to each of their associated data cleaner agents, gather recent information, and classify the data to determine whether suspicious activity is occurring. These agents collaborate to set their suspicion level to determine cooperatively whether a suspicious action is more interesting in the presence of other suspicious activity.
- **High-level agents:** they maintain the data warehouse by combining knowledge and data from the low-level agents. The high-level agents apply data mining algorithms to discover associations and patterns.
- **Interface agent:** it directs the operation of the agents in the system, maintains the status reported by the mobile agents, and provides access to the data warehouse features.

In [48] a multiagent IDS (MAIDS) architecture containing mobile agents is proposed. These lightweight agents, located in the middle of the architecture, form the first line of ID. They periodically travel between monitored systems, obtain the gleaned information, and classify the data to determine whether singular intrusions have occurred.

In the MA-IDS architecture [49] mobile agents are employed to coordinately process information from each monitored host. Only the critical components in the MA-IDS architecture (Assistant and Response agents) are designed as mobile agents. An Assistant mobile agent is dispatched by the Manager component to patrol (gather information) in the network. Assistant mobile agents are intended to determine whether some suspicious activities in different hosts are part of a distributed intrusion. If that is the case, the Manager component will possibly dispatch a Response mobile agent to "intelligently" response to each monitored host. It is claimed that these mobile agents are capable of evading attackers and resurrecting themselves when attacked. Additionally, agent mobility makes distributed ID possible by means of data correlation and cooperative detection.

An interesting and comprehensive discussion about optimising the analysis of NIDSs through mobile agents is presented in [50]. The main proposal is to place the mobile analyser components of the NIDS closer together in the network and shifting the processing load to underused nodes if possible.

APHIDS (Agent-Based Programmable Hybrid Intrusion Detection System) [51] implements the distributed search and analysis tasks with mobile agents equipped with scripting capability to automate evidence gathering. This architecture is similar to the SPARTA and MAIDS systems (all of them exploit the mobility of the agents to perform distributed correlation), but APHIDS allows the specification of coordinated analysis tasks using a high-level specification language. Mobile agents are used for monitoring the output from other previously running IDSs (HIDSs or NIDSs), querying the log files and system state, and reporting results.

APHIDS was subsequently upgraded, generating APHIDS++ [52] that introduces a two-level caching scheme:

- Task Agents enter the first level cache mode (busy wait at the attacked machine) after having handled an initial attack. Each Task Agent maintains a publicly accessible queue of pending attacks to handle.
- If no new attacks are sent to a Task Agent within a certain time limit, the agent enters the second cache level mode, in which it is flushed to its host machine's disk. Thus, resource consumption in the host machine is reduced.

Some other improvements of APHIDS++ are the addition of an optional intelligent agent and an XML implementation of the Distributed Correlation Script.

Two different agent classes are proposed in [53]: monitoring agents (AM) and managing agents (AZ). AM observe the nodes, process captured information, and draw conclusions for the evaluation of the current state of system security. AM agents can travel along the network to monitor different areas that may be at risk of attacks. On the other hand, AZ agents are responsible for creating profiles of attacks, managing AM agents, and updating its database and ontology.

IDReAM (Intrusion Detection and Response executed with Agent Mobility) is proposed in [54] as a new approach to build a completely distributed and decentralized ID and Response System in computer networks. Conceptually, IDReAM combines mobile agents with self-organizing paradigms inspired by natural life systems: immune system that protects the human body against external aggressions and the stigmergic paradigm of a colony of ants. The two natural systems exhibit a social life by the organisation of their entities (immune cells and ants) which the author states is not possible without mobility. IDReAM is assessed in terms of resource consumption and intrusion response efficiency.

IDSUDA (Intrusion Detection System Using Distributed Agents) [55] proposes the application of mobile agents to monitor the usage of various system resources in order to detect deviations from normal usage. The behaviour of the attacker is tracked by following up the intruder movements from one resource to another.

A general distributed IDS framework based on mobile agents is proposed in [56]. Some of the components in such model are designed as mobile agents for the purpose of high adaptability and security of the system. It is claimed that these mobile agents can evade intrusion and recover by themselves if they suffer from intrusion, but further explanations of that are not provided.

## 4 Conclusions

A great effort has been devoted to the MAS approach for Intrusion Detection. Plenty of works have been released on this subject, enabling the ID task in complex and distributed environments. The use of dynamic MASs enables taking advantage of some of the properties of agents such as reactivity, proactivity, and sociability. One of the main weaknesses of such solutions is the defence mechanisms of the MASs, as the resistance to attacks has not been considered in most previous work.

Although mobile agents can provide an IDS with some advantages (mobility, overcoming network latency, robustness, and fault tolerance), some problems have not been completely overcome yet [41]: speed, volume of the code required to implement a mobile agent, deployment, limited methodologies and tools, security threats, and so on.

**Acknowledgments.** This research has been partially supported by the projects BU006A08 of the JCyL and CIT-020000-2008-2 of Spanish Ministry of Science and Innovation. The authors would also like to thank the manufacturer of components for vehicle interiors, Grupo Antolin Ingeniería, S.A. in the framework of the project MAGNO 2008 - 1028.- CENIT Project funded by the Spanish Ministry of Science and Innovation.

## References

1. Chuvakin, A.: Monitoring IDS. Information Security Journal: A Global Perspective 12(6), 12–16 (2004)
2. Frank, J.: Artificial Intelligence and Intrusion Detection: Current and Future Directions. In: 17th National Computer Security Conf., Baltimore, MD, vol. 10 (1994)
3. Mukherjee, B., Heberlein, L.T., Levitt, K.N.: Network Intrusion Detection. IEEE Network 8(3), 26–41 (1994)
4. Engelhardt, D.: Directions for Intrusion Detection and Response: a Survey. Electronics and Surveillance Research Laboratory, Defence Science and Technology Organisation, Department of Defence, Australian Government (1997)
5. Jones, A., Sielken, R.: Computer System Intrusion Detection: A Survey. White paper. University of Virginia - Computer Science Department (1999)
6. Debar, H., Dacier, M., Wespi, A.: Towards a Taxonomy of Intrusion-Detection Systems. Computer Networks - the International Journal of Computer and Telecommunications Networking 31(8), 805–822 (1999)
7. Axelsson, S.: Intrusion Detection Systems: A Survey and Taxonomy. Technical Report. Chalmers University of Technology. Department of Computer Engineering (2000)
8. Allen, J., Christie, A., Fithen, W., McHugh, J., Pickel, J., Stoner, E.: State of the Practice of Intrusion Detection Technologies. Technical Report CMU/SEI-99-TR-028. Carnegie Mellon University - Software Engineering Institute (2000)
9. McHugh, J.: Intrusion and Intrusion Detection. International Journal of Information Security 1(1), 14–35 (2001)
10. Verwoerd, T., Hunt, R.: Intrusion Detection Techniques and Approaches. Computer Communications 25(15), 1356–1365 (2002)

11. Mukkamala, S., Sung, A.H.: A Comparative Study of Techniques for Intrusion Detection. In: 15th IEEE International Conference on Tools with Artificial Intelligence, pp. 570–577 (2003)
12. Estevez-Tapiador, J.M., Garcia-Teodoro, P., Diaz-Verdejo, J.E.: Anomaly Detection Methods in Wired Networks: a Survey and Taxonomy. Computer Communications 27(16), 1569–1584 (2004)
13. Lazarevic, A., Kumar, V., Srivastava, J.: Intrusion Detection: a Survey. In: Managing Cyber Threats: Issues, Approaches, and Challenges 5. Massive Computing, pp. 19–78. Springer, US (2005)
14. Patcha, A., Park, J.-M.: An Overview of Anomaly Detection Techniques: Existing Solutions and Latest Technological Trends. Computer Networks 51(12), 3448–3470 (2007)
15. García-Teodoro, P., Díaz-Verdejo, J., Maciá-Fernández, G., Vázquez, E.: Anomaly-based Network Intrusion Detection: Techniques, Systems and Challenges. Computers & Security 28(1-2), 18–28 (2009)
16. Wooldridge, M., Jennings, N.R.: Agent theories, architectures, and languages: A survey. Intelligent Agents (1995)
17. Franklin, S., Graesser, A.: Is It an Agent, or Just a Program? A Taxonomy for Autonomous Agents. In: Jennings, N.R., Wooldridge, M.J., Müller, J.P. (eds.) ECAI-WS 1996 and ATAL 1996. LNCS, vol. 1193, pp. 21–35. Springer, Heidelberg (1997)
18. Russell, S.J., Norvig, P.: Artificial Intelligence: a Modern Approach. Prentice Hall, Englewood Cliffs (1995)
19. Weiss, G.: Multiagent Systems: a Modern Approach to Distributed Artificial Intelligence. MIT Press, Cambridge (1999)
20. Ferber, J.: Multi-agent Systems: an Introduction to Distributed Artificial Intelligence. Addison-Wesley, Reading (1999)
21. Durfee, E.H., Lesser, V.R.: Negotiating Task Decomposition and Allocation Using Partial Global Planning. In: Distributed Artificial Intelligence, vol. 2. Morgan Kaufmann Publishers Inc., San Francisco (1989)
22. Jennings, N.R., Sycara, K., Wooldridge, M.: A Roadmap of Agent Research and Development. Autonomous Agents and Multi-Agent Systems 1(1), 7–38 (1998)
23. Wooldridge, M.: Agent-based Computing. Interoperable Communication Networks 1(1), 71–97 (1998)
24. Stolfo, S., Prodromidis, A.L., Tselepis, S., Lee, W., Fan, D.W., Chan, P.K.: JAM: Java Agents for Meta-Learning over Distributed Databases. In: Third International Conference on Knowledge Discovery and Data Mining, pp. 74–81 (1997)
25. Reilly, M., Stillman, M.: Open Infrastructure for Scalable Intrusion Detection. In: IEEE Information Technology Conference, pp. 129–133 (1998)
26. Spafford, E.H., Zamboni, D.: Intrusion Detection Using Autonomous Agents. Computer Networks: The International Journal of Computer and Telecommunications Networking 34(4), 547–570 (2000)
27. Hegazy, I.M., Al-Arif, T., Fayed, Z.T., Faheem, H.M.: A Multi-agent Based System for Intrusion Detection. IEEE Potentials 22(4), 28–31 (2003)
28. Gorodetski, V., Kotenko, I., Karsaev, O.: Multi-Agent Technologies for Computer Network Security: Attack Simulation, Intrusion Detection and Intrusion Detection Learning. Computer Systems Science and Engineering 18(4), 191–200 (2003)
29. Miller, P., Inoue, A.: Collaborative Intrusion Detection System. In: 22nd International Conference of the North American Fuzzy Information Processing Society (NAFIPS 2003), pp. 519–524 (2003)

30. Gorodetsky, V., Karsaev, O., Samoilov, V., Ulanov, A.: Asynchronous alert correlation in multi-agent intrusion detection systems. In: Gorodetsky, V., Kotenko, I., Skormin, V.A. (eds.) MMM-ACNS 2005. LNCS, vol. 3685, pp. 366–379. Springer, Heidelberg (2005)
31. Dasgupta, D., Gonzalez, F., Yallapu, K., Gomez, J., Yarramsettii, R.: CIDS: An agent-based intrusion detection system. Computers & Security 24(5), 387–398 (2005)
32. Cougaar: Cognitive Agent Architecture, http://cougaar.org/
33. Debar, H., Curry, D., Feinstein, B.: The Intrusion Detection Message Exchange Format (IDMEF). IETF RFC 4765 (2007)
34. Gowadia, V., Farkas, C., Valtorta, M.: PAID: A Probabilistic Agent-Based Intrusion Detection system. Computers & Security 24(7), 529–545 (2005)
35. Tsang, C.-H., Kwong, S.: Multi-agent Intrusion Detection System in Industrial Network using Ant Colony Clustering Approach and Unsupervised Feature Extraction. In: IEEE International Conference on Industrial Technology (ICIT 2005), pp. 51–56 (2005)
36. Mukkamala, S., Sung, A.H., Abraham, A.: Hybrid Multi-agent Framework for Detection of Stealthy Probes. Applied Soft Computing 7(3), 631–641 (2007)
37. Herrero, Á., Corchado, E., Pellicer, M.A., Abraham, A.: MOVIH-IDS: A Mobile-Visualization Hybrid Intrusion Detection System. Neurocomputing 72(13-15), 2775–2784 (2009)
38. Corchado, J.M., Laza, R.: Constructing Deliberative Agents with Case-Based Reasoning Technology. International Journal of Intelligent Systems 18(12), 1227–1241 (2003)
39. Pellicer, M.A., Corchado, J.M.: Development of CBR-BDI Agents. International Journal of Computer Science and Applications 2(1), 25–32 (2005)
40. Aamodt, A., Plaza, E.: Case-Based Reasoning - Foundational Issues, Methodological Variations, and System Approaches. AI Communications 7(1), 39–59 (1994)
41. Jansen, W.A., Karygiannis, T., Marks, D.G.: Applying Mobile Agents to Intrusion Detection and Response. US Department of Commerce, Technology Administration, National Institute of Standards and Technology (1999)
42. Asaka, M., Taguchi, A., Goto, S.: The Implementation of IDA: An Intrusion Detection Agent System. In: 11th Annual Computer Security Incident Handling Conference, vol. 6 (1999)
43. De Queiroz, J.D., da Costa Carmo, L.F.R., Pirmez, L.: Micael: An Autonomous Mobile Agent System to Protect New Generation Networked Applications. In: Second International Workshop on Recent Advances in Intrusion Detection, RAID 1999 (1999)
44. Mell, P., Marks, D., McLarnon, M.: A Denial-of-service Resistant Intrusion Detection Architecture. Computer Networks: The International Journal of Computer and Telecommunications Networking 34(4), 641–658 (2000)
45. Krügel, C., Toth, T., Kirda, E.: SPARTA: a Mobile Agent Based Instrusion Detection System. In: IFIP TC11 WG11.4 First Annual Working Conference on Network Security: Advances in Network and Distributed Systems Security. IFIP Conference Proceedings, vol. 206, pp. 187–200. Kluwer, Dordrecht (2001)
46. Dasgupta, D., Brian, H.: Mobile Security Agents for Network Traffic Analysis. In: DARPA Information Survivability Conference & Exposition II (DISCEX 2001), vol. 2, pp. 332–340 (2001)
47. Helmer, G., Wong, J.S.K., Honavar, V.G., Miller, L.: Automated Discovery of Concise Predictive Rules for Intrusion Detection. Journal of Systems and Software 60(3), 165–175 (2002)
48. Helmer, G., Wong, J.S.K., Honavar, V., Miller, L., Wang, Y.: Lightweight Agents for Intrusion Detection. Journal of Systems and Software 67(2), 109–122 (2003)

49. Li, C., Song, Q., Zhang, C.: MA-IDS Architecture for Distributed Intrusion Detection using Mobile Agents. In: 2nd International Conference on Information Technology for Application (ICITA 2004), pp. 451–455 (2004)
50. Marks, D.G., Mell, P., Stinson, M.: Optimizing the Scalability of Network Intrusion Detection Systems Using Mobile Agents. Journal of Network and Systems Management 12(1), 95–110 (2004)
51. Deeter, K., Singh, K., Wilson, S., Filipozzi, L., Vuong, S.T.: APHIDS: A mobile agent-based programmable hybrid intrusion detection system. In: Karmouch, A., Korba, L., Madeira, E.R.M. (eds.) MATA 2004. LNCS, vol. 3284, pp. 244–253. Springer, Heidelberg (2004)
52. Alam, M.S., Gupta, A., Wires, J., Vuong, S.T.: APHIDS++: Evolution of A programmable hybrid intrusion detection system. In: Magedanz, T., Karmouch, A., Pierre, S., Venieris, I.S. (eds.) MATA 2005. LNCS, vol. 3744, pp. 22–31. Springer, Heidelberg (2005)
53. Kolaczek, G., Pieczynska-Kuchtiak, A., Juszczyszyn, K., Grzech, A., Katarzyniak, R.P., Nguyen, N.T.: A mobile agent approach to intrusion detection in network systems. In: Khosla, R., Howlett, R.J., Jain, L.C. (eds.) KES 2005. LNCS (LNAI), vol. 3682, pp. 514–519. Springer, Heidelberg (2005)
54. Foukia, N.: IDReAM: Intrusion Detection and Response Executed with Agent Mobility Architecture and Implementation. In: Fourth International Joint Conference on Autonomous Agents and Multiagent Systems (AAMAS 2005). ACM, The Netherlands (2005)
55. Alim, A.S.A., Ismail, A.S., Ahmed, S.H.: IDSUDA: An Intrusion Detection System Using Distributed Agents. Journal of Computer Networks and Internet Research 5(1), 1–11 (2005)
56. Wang, H.Q., Wang, Z.Q., Zhao, Q., Wang, G.F., Zheng, R.J., Liu, D.X.: Mobile agents for network intrusion resistance. In: Shen, H.T., Li, J., Li, M., Ni, J., Wang, W. (eds.) APWeb Workshops 2006. LNCS, vol. 3842, pp. 965–970. Springer, Heidelberg (2006)

# Multimodal Biometrics: Topics in Score Fusion

Luis Puente, M. Jesús Poza, Juan Miguel Gómez, and Diego Carrero

Universidad Carlos III de Madrid. Avda Universidad, 30. 28911 Leganés, Madrid, Spain
Tel.: (+34) 91 624 99 68; Fax: (+34) 91 624 91 29
lpuente@it.uc3m.es, mpoza@pa.uc3m.es,
juanmiguel.gomez@uc3m.es, dcarrero@di.uc3m.es

**Abstract.** This paper describes how the last cutting-edge advances in Computational Intelligence are being applied to the field of biometric security. We analyze multimodal identification systems and particularly the score fusion technique and some issues related with it. Fundamentally, the paper deals with the scores normalization problem in depth, which is one of the most critical issues with a dramatic impact on the final performance of multibiometric systems. Authors show in this paper the results obtained using a number of fusion algorithms (Neural Networks, SVM, Weighted Sum, etc.) on the scores generated with three independent monomodal biometric systems (the modalities are Iris, Signature and Voice). The paper shows the behavior of the most popular score normalization techniques (z-norm, tanh, MAD, etc), and proposes a new score normalization procedure with an optimized performance harnessing tested fusion techniques and outperforming previous results through a proof-of-concept implementation.

**Keywords:** Score Normalization, Multi-Modal biometrics, Score Fusion.

## 1 Introduction

Computational intelligence (CI) is a research field with approaches primarily based on learning machines. Computational intelligence technologies are being successfully applied to complex problems, and they have been proved to be effective and efficient in biometric matching tasks, sometimes used in combination with traditional methods. Using these technologies, biometrics is nowadays rapidly evolving; it becomes more and more attractive and effective in security applications, such as user authentication to control the access to personal information or physical areas.

The goal of a biometrical authentication system is to validate the persons' identity based on one or more previously chosen particular physical or behavioral traits (voice, image of the iris, fingerprint, etc.) [1].

Biometrical traits used in authentication systems should satisfy several requirements [2], namely: Universality, Permanence, Collectability, Acceptability and Distinctiveness. None of the used biometrical traits used fulfill all these requirements hence this is one of the reasons that make unimodal biometric authentication systems fail in some circumstances, even in friendly environments. Therefore, within the aim of improving biometric authentication systems, we focus on increasing the recognition rates of traditional systems and improving their usability by reducing strong dependence on environmental conditions and the availability of the biometric traits. This usability gain will be achieved by making systems that are based on various biometrical features. Particularly since we are not considering just one, but a whole lattice of them, having a multimodal biometric system to rely on. Multimodality will improve the performance of authentication systems. It will provide robustness against individual sensor or subsystem failures and, finally, will reduce the occasions in which the system cannot make a decision.

In this paper, the fusion of three unimodal systems into a multimodal one was carried out at a "score" level [3]: each unimodal system generates a verisimilitude (score), and the decision of the multimodal system will come from the fusion of the three different unimodal scores. The paper will show the results and conclusions obtained during the process carried out to determine which normalization technique and which fusion method achieve the best results in people's identity verification based on three previously selected unimodal processes. The biometric modalities used were handwritten signatures, irises and speech. The authors propose a new score normalization algorithm (SDN) that provides a light improvement on the performance of the final decisor if we compare it with the traditional normalization techniques. The rest of the work is structured in the following way: in Sections 2, 3 and 4 state of the art biometric fusion is revised. Section 5 presents the characteristics of the data used in the tests. and shows the results achieved. In Section 6, the authors compile their conclusions.

## 2 Taxonomy of Biometric Systems

### 2.1 Unimodal Systems

The first step taken by a biometric verification system is to obtain a set of data from a biometric sample presented by a person (donor) to the respective sensor. The immediate aim of the system is using this data frame to determine the probability (confidence) that the sample belongs to the user that the donor claims to be (claimed user). This is achieved by "comparing" the data frame with the claimed user's model. For each user of the system, his/her model is calculated in a previous phase named the training or enrollment phase. Furthermore, the processing of biometric samples is carried out, and in this phase one must be sure that the biometric samples belong to the user whose model is being generated.

The biometric systems present some limitations in their use [4]: noise, intra-class and inter-class variations, non-universality... all these limitations prevent the unimodal verification systems from fulfilling the expectancy of a security system, and suggest looking for choices that provide higher usability, robustness and performance.

### 2.2 Multimodal Systems

The limitations of the unimodal biometric systems can make biometric verification become the weak point of the security of the authentication system. In order to partially or totally solve these defects, diverse alternatives have been proposed for joining different techniques into a single system [4][5], which has been called Biometric Fusion. According to this, a whole lattice of fusion scenarios have been proposed:

multiple sensors/capture units, multiple takes (a single sensor), multiple instances of a single biometric modality but using multiple parts of the human body, multiple algorithms, multiple traits...both multiple sensors and multiple takes belong to the pre-classification category of fusion strategies: the information is fused before using any pattern matching algorithm. In multiple algorithms and multiple traits scenarios, the information fusion can be made at feature level (pre-classification) or at score or decision level (post-classification strategy).

A multibiometric system can be based on one or a combination of several of these scenarios. Multibiometric systems integrate the evidence presented by multiple biometric sources and are more robust to variations in sample quality than unimodal systems due to the presence of multiple (and usually independent) pieces of evidence. The last group of techniques (multiple traits) is called multimodal fusion: combining different biometric modalities is a good way to enhance biometric systems, given that different traits fulfill the well known requirements of Universality, Permanence, Distinctiveness, Circumvention and Collectability [2].

Multimodal fusion (a combination of various biometric modalities) is the most popular technique due to the quality of its results. In almost all cases, it is based on carrying out a group of unimodal processes obtaining a decision and a confidence value (scores) in each of them. By means of the appropriate combination of these unimodal results, it is possible to express a final decision about the donor's identity. To carry out this fusion, different criteria can be used, using either the unimodal decisions or the unimodal scores, and no agreement has being reached on what the best solution is because different environments, different sets of traits, different sets of sensors, etc. seem to require unique particular processes.

## 3 Score Fusion

Score level fusion refers to the combination of matching scores provided by the unimodal classifiers in the system. This is the most widely used fusion approach, as evidenced by the experts in the field. A priori, one could think that merging information from the different modalities at some previous stage of the system (sensor level, feature level) will provide more effectiveness, but there are several reasons that support score fusion, such as conceptual simplicity, ease implementation, practical aspects, etc. Thus, the dominant option in the majority of published papers is score-level fusion.

As it has been mentioned, there are two categories of score fusion: classification (or learning based) fusion, where matching scores are considered 'input features' for a new pattern-classification problem, and combination (or rule-based) fusion, where the input matching scores are normalized and combined to obtain a final score [4]. One can expect that the classification techniques behave better than the combination techniques (comparing error rates), but the score combination algorithms are easier to implement and have lower computational costs than the classification algorithms.

Despite the method used in score fusion, it should be taken into account that the scores produced by the different unimodal systems may not be homogeneous. They are probably not in the same numerical scale or follow the same statistical distribution and, in addition, they are usually not comparable, given that the output of a biometric system can be a probability, or a measure of distance, a similarity, or a log-likelihood ratio. Moreover, decisions are often made by comparing the classifier score with a threshold. This makes the score normalization step an important part in the design of a multimodal system, as scores must be made comparable by using score normalization techniques.

The authors of this paper show the results obtained in their process of selecting the best score fusion method and the best score normalization method for the particular problem of verifying identity in based on signature, iris and voice traits. The authors' idea of score fusion includes score combination and score classification. Four classifiers were chosen: Weighted Sum [6], Weighted Product [7], Neural Networks (NN) [8], and Support Vector Machines (SVM) with the linear, polynomial and RBF kernels [9].

## 4 Score Normalization

Score normalization refers to changing the location and scale parameters of the matching score distributions at the outputs of the unimodal authentication systems, so that the matching scores of different matchers are transformed into a common domain. The significance of this step has already been outlined in the paper.

There are several normalization algorithms in the literature: from the simplest ones (e.g., min-max) to the histogram equalization or sigmoid functions, such as tanh(). The authors have tested the min-max, the z-score, the MAD, the tanh, the atanh and the double sigmoid algorithms for score normalization [5].

Besides the aforementioned score normalization techniques, the authors of this paper propose a new one: the Simple Data Normalization (SDN) algorithm, which shows utility when the decision thresholds in each modality are different from each other. Precisely, this is the case in this study: in the three modalities they have been used, scores in the range (0,1) and authors observed that the distances between positive scores (the ones that are over the decision threshold) and between negative scores are not equivalent: 'positive' scores are more crowded together than 'negative' scores. This data configuration makes multimodal decision difficult. In order to help the decision task, authors propose to modify the original unimodal scores, so that the threshold in each modality become 0,5. This makes that separation of the positive scores is increased, and the negative and positive decision ranges become equivalent in each modality. This normalization algorithm uses a double linear transformation as can be seen in the formula:

$$x = \begin{cases} \dfrac{s}{2 * th} & s \leq th \\ 0{,}5 + \dfrac{s - th}{2 * (1 - th)} & s > th \end{cases} \quad (1)$$

Once the data has been transformed using this algorithm, authors go one step beyond, and transform again the data, giving more weight to the scores that are further away from the decision threshold: each SDN score turns into twice the square of its distance to the threshold, maintaining the sign. We have called this algorithm Cuadratic-SDN. The formula of the new normalized score (x') is

$$x' = 2*(x-0,5)*|x-0,5|\ . \tag{2}$$

## 5 Experimental Methodology and Results

The starting point of tests was the group of results obtained by three independent unimodal authentication processes: Iris Image Recognition (iris), Handwritten Signature verification (signature) [10] and Speaker Verification (voice). Each unimodal system uses a different biometric data base and the main parameters and performances are different too:

- The voice system runs on the BIOSEC [11] data base, uses a SV-Machine with a decision threshold of 67%, the result was an EER of 25.0%.
- The data of the signature modality come from the MCYT Signature Corpus Database [12]; [10] describes the authentication system. The resulting system EER was 5.73% using a decision threshold of 91%.
- The biometric data of irises was obtained from different public databases [13][14][15]. The system is described in [16]. Using a predefined decision threshold (65%), the resulting equal error rate was 12.4%.

In this paper, experimental results are expressed in terms of the EER (Equal Error Rate), i.e., the crossing point of the false acceptance and the false rejection curves.

In our tests, the outputs of individual modalities were only matching scores without any measures quantifying the confidence in those scores. The three unimodal outputs were not homogeneous (some are probabilities while others are metrical distances), and –of course- they did not belong to the same numerical range. Thus, the first step was to transform them. Anyway and due to the diversity of the sources of the samples, it seems clear that the results obtained at the end of the unimodal processes are uncorrelated: in order to obtain the input data for the final classifier, these results (after normalizing them) were mixed randomly to conform tuples without any restriction, except:

- In each tuple, one score of each modality
- Not mixing results of genuine samples (the donor is the claimed user) with others of non-genuine samples (recordings where the donor is a person different from the claimed user) in the same tuple.

As a result of the above criterion, a set of 5684 full-genuine tuples and a set of 6000 full-non-genuine tuples were obtained. The information contained in each of these tuples is *"Iris confidence," "Signature confidence," "Voice confidence" and "Genuine or not."* One thousand tuples were randomly selected to compound the training set: 500 from the full-genuine and 500 from the full-non-genuine set. The rest of the tuples were used as a test set. These two groups were the ones used in the rest of the experiments performed.

There were seven algorithms to be analyzed: four combination algorithms (Voting, Minimum, Maximum, Weighted Sum and Weighted Product) and two classification algorithms (SVM –with different kernels- and Multi-layer Neural Network), although only the best results will be depicted in this paper (so, voting, maximum and minimum don't appear).

For the calculation of the weighted sum coefficients, a monolayer perceptron was used. For the case of the weighted product, logarithms were used to turn it into a weighted sum so that we could use the same type of perceptron in the determination of the coefficients.

The tests we have performed are been designed to gain knowledge about the behaviour of the several normalization algorithms and the behaviour of the before mentioned fusion algorithms. The results (in EER terms) are depicted in Table 1, where columns stand for normalization algorithms (DS means double sigmoid, SDN_S is a sigmoid normalization applied on data with a SDN normalization and Q_SDN is quadratic SDN) and rows show results of fusion algorithms.

Table 1. EER (%) obtained in the experiments

|  | min_max | tanh | atanh | z_score | MAD | DS. | SDN | SDN_S | Q_DSN |
|---|---|---|---|---|---|---|---|---|---|
| NN(3-3-1) | 2,73 | 2,66 | 2,03 | 1,81 | **1,66** | 4,64 | **1,66** | 1,75 | 2,23 |
| SVM_LINEAR | 2,28 | 3,83 | 1,73 | 2,26 | 2,26 | 4,55 | 1,71 | 1,85 | 2,19 |
| SVM_POLIN. | 2,63 | 3,31 | 1,94 | 2,28 | 2,51 | 4,11 | 1,94 | 2,10 | 2,66 |
| SVM_RADIAL | 2,28 | 3,86 | 1,73 | 1,73 | 1,75 | 4,45 | 1,71 | 1,88 | 2,28 |
| SVM_SIGMOID | 2,47 | 3,80 | 1,75 | 4,45 | 5,03 | 4,51 | 1,73 | 1,94 | 2,34 |
| Weighted Pr. | 2,48 | 12,84 | 1,73 | 9,84 | 9,80 | 4,98 | 1,78 | **1,66** | 2,03 |
| Weighted Sum | 2,28 | 3,51 | 1,73 | 2,31 | 2,31 | 4,93 | 1,73 | 1,75 | 2,06 |

Table 2 presents the ranking of the best explored techniques (it exhibits EER, Area Under ROC Curve and the NIST Cost function [17]). The Simple Data Normalization presented in this paper has proved being simple in implementation and effective in the results improvement. Using other functions, such as the sigmoid, a higher resolution

Table 2. Ranking

| Fusion-normalization method | EER | AUC | COST |
|---|---|---|---|
| NN (3-3-1) SDN | 1,66% | 99,86% | 3,20% |
| NN(3-3-1) MAD | 1,66% | 99,85% | 2,70% |
| Weighted Pr SDN-S | 1,66% | 99,86% | 2,03% |
| SVM (LINEAR, RBF) SDN | 1,71% | 99,84% | 2,34% |
| WP ATANH | 1,73% | 99,85% | 2,23% |
| WS ATANH | 1,73% | 99,86% | 2,22% |
| SVM(LINEAR, RBF) ATANH | 1,73% | 99,86% | 2,30% |
| NN(3-3-1) SDN | 1,73% | 99,86% | 3,20% |

around the threshold is achieved, increasing the indecision of the multimodal fusion, as this gives more weight to results classified as 'low confidence' by the unimodal classifiers.

## 6 Conclusions

The main conclusion is that multimodal fusion outperforms considerably unimodal systems: from data with low quality EER (12.4%, 5.7% and 25,0%) the fusion algorithms have managed to obtain noticeably improved results. One of the straightforward consequences of this conclusion is that we can use algorithms of unimodal classification computationally economic, mainly because the multimodal fusion of its results generates outputs corresponding to high facilities systems.

This improvement affects all techniques previously described. Taking into account some studies comparing normalization techniques and fusion algorithms, one can notice that the improvement depends upon the used algorithms as well as the particular situation we confront. Therefore, a priori we cannot name the best techniques for a particular case. It would be necessary to test several of them so that it could be decided which normalization and which fusion method is the best.

Furthermore, multimodal fusion permits to authenticate a greater number of people since more information sources are available. If an individual is incapable of producing a particular biometrical trait, he could, nonetheless, generate other characteristics for which he could be recognized.

To sum up, a proper data treatment prior to the fusion step is crucial if the fusion algorithm belongs to the combination category: with an adequate score processing, all fusion algorithms used in our tests behave in a very similar manner, despite its computational complexity is very different.

**Acknowledgments.** This work has been partially developed within the PIBES (TEC2006-12365-C02-01) and the BRAVO (TIN2007-67407-C03-01) Projects founded by the Spanish Ministry of Science and Innovation.

## References

1. Furui, S.: Recent advances in speaker recognition. In: Audio- and Video-based Biometric Person Authentication, pp. 237–252. Springer, Heidelberg (1997)
2. Hong, L., Jain, A.K., Pankanti, S.: Can Multibiometrics Improve Performance? In: Proceedings AutoID 1999, Summit, NJ, October 1999, pp. 59–64 (1999)
3. Sigüenza, A., Tapiador, M.: Tecnologías Biométricas Aplicadas a la Seguridad. Editorial Ra-Ma (2005)
4. Jain, A.K., Ross, A., Prabhakar, S.: An Introduction to Biometric Recognition. IEEE Trans. Circuits Syst. Video Technology, Special Issue Image- and Video-Based Biomet. 14(1), 4–20 (2004)
5. Ross, A., Jain, A.K.: Information Fusion in Biometrics. Pattern Recognition 24(13), 2115–2125 (2003)
6. Huang, Y., Suen, C.: A Method of Combining Multiple Experts for the Recognition of Unconstrained Handwritten Numerals. IEEE Trans. on PAMI 17, 90–94 (1995)

7. Kittler, J., Hatef, M., Duin, R., Matas, J.: On Combining Classifiers. IEEE Transacations on Pattern Analysis and Machine Intelligence 20(3), 226–239 (1998)
8. Bishop, C.M.: Neural Networks for Pattern Recognition. Oxford University Press, Oxford (1995)
9. Vapnic, V.N.: The Nature of Statistical Learning Theory. Springer, Heidelberg (1995)
10. Miguel-Hurtado, O., Mengibar-Pozo, L., Lorenz, M.G., Liu-Jimenez, J.: On-Line Signature Verification by Dynamic Time Warping and Gaussian Mixture Models. In: IEEE Proc. Int. Carnahan Conference on Security Technology, October 2007, pp. 23–29 (2007)
11. Sanchez-Avila, C., Sanchez-Reillo, R.: Two Different Approaches for Iris Recognition using Gabor Filters and Multiscale Zero-crossing Representation. Pattern Recognition 38, 231–240 (2005)
12. Ortega-García, J., et al.: MCYT baseline corpus: a bimodal biometric database. IEE Proceedings Vision, Image and Signal Processing 150(6), 395–401 (2003)
13. CASIA Iris Data base, http://www.cbsr.ia.ac.cn/Databases.htm
14. Univerzita Palackého Iris Database, http://www.inf.upol.cz/iris/
15. NIST ICE2005 Iris Database, http://iris.nist.gov/ICE/
16. Fernandez-Saavedra, B., Liu-Jimenez, J., Sanchez-Avila, C.: Quality Measurements for Iris Images for Biometrics. In: IEEE EUROCON 2007 International Conference on "Computer as a tool", Warsaw, Poland (September 2007)
17. Przybocki, M.A., Martin, A.F., Le, A.N.: NIST speaker recognition evaluations utilizing the mixer corpora-2004,2005,2006. IEEE trans. on Audio, Speech and Language Processing 15(7) (September 2007)

# Security Efficiency Analysis of a Biometric Fuzzy Extractor for Iris Templates

F. Hernández Álvarez and L. Hernández Encinas

Department of Information Processing and Coding
Applied Physics Institute, CSIC
C/ Serrano 144, 28006-Madrid, Spain
{fernando.hernandez,luis}@iec.csic.es

**Abstract.** A Biometric fuzzy extractor scheme for iris templates was recently presented in [3]. This fuzzy extractor binds a cryptographic key with the iris template of a user, allowing to recover such cryptographic key by authenticating the user by means of a new iris template from her. In this work, an analysis of the security efficiency of this fuzzy extractor is carried out by means of a study about the behavior of the scheme with cryptographic keys of different bitlengths: 64, 128, 192, and 256. The different sizes of the keys permit to analyze the variability of the intra- and inter-user in the iris templates.

**Keywords:** Biometrics, Iris Template, False Acceptance Rate, False Rejection Rate, Fuzzy extractor.

## 1 Introduction

One of the most important uses of Biometrics nowadays is to identify and authenticate individuals by means of one or several of their physiological or behavioral features, like fingerprint, voice, hand geometry, iris, retina, signature, etc.

In general, to validate the identity of a user, biometric procedures consists into two phases: The enrollment and the verification phase. In the enrollment phase, the biometric templates are processed and stored in the database. In the verification phase, a new biometric template (called the query template) is extracted from the user ant it is compared with the data already stored. If the comparison matches, the user identity is validated.

The main properties which permit to consider biometric data as good candidates for security applications are the following: They are unique to each user, they are hard to forge, and they have a good source of entropy. Nevertheless, biometric templates present some drawbacks. Among them the most important are the intra- and inter-user variability.

The intra-user variability measures the differences of two biometric templates extracted from the same user, while the inter-user variability measures the similarities between two biometric templates extracted from different users. These two measurements can cause not to recognize a known user or to recognize an attacker as a known user. The ratios used to measure these two subjects are, respectively: False Rejection Rate ($FRR$) and False Acceptance Ratio ($FAR$).

One desirable characteristic that the biometric templates should have is to be revoked or canceled if necessary, as PIN and passwords do. Several approaches, known as biometric template protection schemes ([6]), have been proposed to secure biometric templates and they can be broadly classified into two categories, namely, feature transformation approach and biometric cryptosystem.

In the feature transformation approach a transformation function is applied to the biometric template before storing it in the database. The function used can be invertible or non-invertible. On the other hand, biometric cryptosystems ([10]) need to generate public information, known as helper data, about the biometric template in order to perform the verification phase. These systems can be classified into three models, key release, key binding and key generation.

In a key binding cryptosystem, the biometric template is secured by binding it with a key in a cryptographic framework. The key and the biometric template are stored in the database as a single entity which represents the helper data. This system has the advantage that it is tolerant to intra-user variability, but this tolerance is determined by the error correcting capability. The limitation of this system is that the matching has to be done using error correction schemes and therefore it is necessary the use of sophisticated matchers. Another limitation is the way the helper data are designed. Several template protection technologies can be considered as key binding approaches, for example fuzzy commitment scheme ([5]), fuzzy vault scheme ([4]), etc.

The fuzzy vault scheme is a cryptographic framework that binds a biometric template with a secret key to build a secure sketch of this template. This sketch is the data which are stored because it is computationally hard to retrieve either the template or the key without any knowledge of the user's biometric data.

In this work, an analysis of the security efficiency of the fuzzy extractor scheme for iris templates proposed in [3] is carried out. This is done by means of a complete study about the behavior of the scheme with cryptographic keys of different bitlengths: 64, 128, 192, and 256. The different sizes of the keys will permit to analyze the intra- and inter-user variability of the iris templates.

The rest of this work is organized as follows. In Section 2 a short review of the fuzzy extractor scheme proposed in [3] is presented. The security efficiency analysis of that scheme is carried out in Section 3; and finally, the conclusions of this work are presented in Section 4.

## 2 Review of a Biometric Fuzzy Extractor for Iris Templates

In [3] a biometric fuzzy extractor scheme for iris templates was presented. It is based on fuzzy extractor schemes ([2]) and on fuzzy vault schemes ([4]).

As it is well-known, fuzzy extractor's basic aim, according to the definitions of Dodis *et al.* ([2]), is to authenticate a user using her own biometric template, $\mathcal{B}$, as the key. To do so, it makes use of another process known as secure sketch to allow precise reconstruction of a noisy input. The correctness of the whole procedure depends on the Hamming distance between $\mathfrak{B}$, used in the enrollment phase, and the query template $\bar{\mathfrak{B}}$, used in the verification phase. Moreover, a fuzzy vault scheme binds a biometric template $\mathfrak{B}$, with a secret (or a key), $S$, to build a secure sketch of $\mathfrak{B}$ itself.

Among the papers related to key binding, the scheme whose efficiency is being measured in this work ([3]) has some similarities with the schemes proposed by Lee et al. ([7]) and Tong et al. ([9]). The main differences among them are that [7] generates the IrisCode from a set of iris features by clustering, technique that is not used in our scheme, and [9] uses fingerprints instead of iris templates.

Therefore, some stages have been adjusted to provide a higher level of security and making them fit with iris templates instead of fingerprints. Another improvement that has been done is that we implement in Java the whole scheme to present some conclusions and useful results.

The two phases associated to this scheme are the following:

## 2.1 Enrollment Phase

In the enrollment phase, from the user's iris template, $\mathfrak{B}$, and a key/secret, $S$, chosen by herself, the scheme produces two sets, $H$ and $\Delta$, as public helper data, which are stored in the database. The different stages of this phase are:

1. The key $S$ is represented in a *base* (10, 16, 256, 512, etc.). The digits of $S$ in that base are considered as the coefficients of a polynomial $p(x)$ of degree $d$. That is, if $S = \{s_0, s_1, \ldots, s_d\}$, then $p(x) = s_0 + s_1 x + s_2 x^2 + \ldots + s_d x^d$.
2. Next, $n$ random integer numbers, $x_i \in \mathbb{Z}$, are generated in order to compute $n$ pairs of points, $(x_i, y_i)$, verifying $p(x)$, i.e., $y_i = p(x_i)$, $0 \leq i \leq n-1$. The parameter $n$ determines the level of fuzziness of the system, so $n$ must be much greater than $d$, $(n \gg d)$.
3. The points are encoded by using a Reed-Solomon code into $n$ codewords determining a set $C = \{c_0, c_1, \ldots, c_{n-1}\}$. This codification is done to avoid somehow the intra-user variability thanks to the error-correction properties of the Reed-Solomon codes.
4. A hash function, $\mathfrak{h}$, is applied to the elements of $C$ to obtain a new set $H = \{\mathfrak{h}(c_0), \mathfrak{h}(c_1), \ldots, \mathfrak{h}(c_{n-1})\}$.
5. The iris template of each user is divided into $n$ parts, as many as points were calculated: $\mathfrak{B} = b_0 \parallel b_1 \parallel \ldots \parallel b_{n-1}$.
6. Then, from the values $b_i$ and $c_i$, $0 \leq i \leq n-1$, the elements of the set $\Delta = \{\delta_0, \delta_1, \ldots, \delta_{n-1}\}$ are calculated, where $\delta_i = c_i - b_i, 0 \leq i \leq n-1$.

Finally, once the helper data ($H$ and $\Delta$) are determined, they are stored in the database. Moreover, the control parameters are made public.

## 2.2 Verification Phase

The first task of this phase is to obtain the control parameters previously stored in the enrollment phase. Then, the following stages are performed:

1. The query iris template, $\tilde{\mathfrak{B}}$, is divided into $n$ parts, as it was done in the enrollment phase: $\tilde{\mathfrak{B}} = \bar{b}_0 \| \bar{b}_1 \| \ldots \| \bar{b}_{n-1}$.
2. Next, from the sets $\Delta$ and $\tilde{\mathfrak{B}}$, a new set is computed: $\bar{C} = \{\bar{c}_0, \bar{c}_1, \ldots, \bar{c}_{n-1}\}$, where $\bar{c}_i = \delta_i + \bar{b}_i$.
   Note that each value $\bar{c}_i$ is supposed to be similar to the corresponding value $c_i \in C$, but with some differences due to the intra-user variability.
3. The same hash function, $\mathfrak{h}$, is applied to the elements of the $\bar{C}$, and the result is compared with the elements of the set $H$.
   In this comparison, at least $d + 1$ coincidences between $H$ and $\mathfrak{h}(\bar{C})$ are necessary to continue with the process, due to the fact that Lagrange interpolation method is used to rebuild the polynomial $p(x)$ of degree $d$. This comparison shows the importance of the parameter $n$, because it will determine the rate of errors admitted in the system due to the intra-user variability.
4. The coincident values are decoded by means of the Reed-Solomon code and $d + 1$ points, $(x_i, y_i)$, at least, are obtained.
5. By using that points and the Lagrange interpolation method, $p(x)$ is rebuilt.
6. Finally, from the coefficients of $p(x)$, the secret $S$ is determined and retrieved to the user.

## 3  Security Efficiency Analysis

In order to determine the security efficiency of the fuzzy extractor scheme presented above, the False Rejection Rate and the False Acceptance Rate of this biometric system will be computed by using different sizes of the secret $S$.

A fix value of 192 bits for $S$ was considered in [3]. In the present analysis different bitlengths of $S$, denoted by $|S|$, will be used to make a comparison between all of them and to determine the security efficiency of the system. The different bitlengths selected are 64, 128, 192, and 256 because they are the standard sizes for cryptographic symmetric keys used nowadays ([8]). This analysis is relevant because if a base to represent the secret $S$ is fixed, the value of the degree $d$ of $p(x)$ depends on the size (bitlength) of $S$, $|S|$.

In this way, when the comparison is carried out in the verification phase, as the value of $d$ is different, a different number of coincidences between $H$ and $\mathfrak{h}(\bar{C})$ are necessary to validate the user's iris template. This fact can be seen as an advantage but at the same time as a drawback, because it can be "easier" to recognize a known user but it does the same to a possible attacker.

The results of this analysis have been obtained by using the same 25 users as those ones used in [3]. These data have been taken from the CASIA database of iris images. Each one of these users have 7 different images of their irises and the corresponding templates of all these $25 \cdot 7 = 175$ images have been extracted by using the algorithm designed by Diez Laiz ([1]).

The parameters used in this analysis are the same (or similar for the Reed-Solomon codes) than those used in [3] in order to do a trustworthy comparison. The only different parameter is the degree of $p(x)$, $d$, as it depends on the bitlength of $S$. In fact, the values considered are: The base used for $S$ is 512; the hash function is $\mathfrak{h} =$ SHA-512; and the fuzziness parameter is $n = 384$.

Security Efficiency Analysis of a Biometric Fuzzy Extractor   167

The values for $d$ as function of the bitlength of $S$ are shown in Table 1.

Table 1. Values of $d$ depending on the bitlength of $S$

| bitlength of $S$: $|S|$ | 64 | 128 | 192 | 256 |
|---|---|---|---|---|
| Value of $d$ | 8 | 14 | 21 | 28 |

## 3.1 Intra-user Variability: FRR

In this analysis each one of the 7 templates of the 25 users is compared with the rest of the templates of the same user. In this way, it is analyzed whether the user is recognized or not, and the False Rejection Rate is determined. The number of comparisons done for each user is $\binom{7}{2} = 21$.

Tables 2, 3, 4, and 5 shows the comparisons obtained with $d+1$ coincidences, at least, for each of the 25 users compared to herself, and for each value of $d$.

Table 2. Number of comparisons with, at least, $d+1 = 8$ coincidences for $|S| = 64$

| | User 1 | User 2 | User 3 | User 4 | User 5 | User 6 | User 7 | User 8 | User 9 |
|---|---|---|---|---|---|---|---|---|---|
| $> d = 7$ | 21 | 21 | 21 | 21 | 21 | 21 | 21 | 15 | 21 |
| | User 10 | User 11 | User 12 | User 13 | User 14 | User 15 | User 16 | User 17 | User 18 |
| $> d = 7$ | 21 | 21 | 21 | 20 | 21 | 21 | 21 | 21 | 21 |
| | User 19 | User 20 | User 21 | User 22 | User 23 | User 24 | User 25 | | |
| $> d = 7$ | 21 | 21 | 21 | 21 | 21 | 21 | 21 | | |

Table 3. Number of comparisons with, at least, $d+1 = 15$ coincidences for $|S| = 128$

| | User 1 | User 2 | User 3 | User 4 | User 5 | User 6 | User 7 | User 8 | User 9 |
|---|---|---|---|---|---|---|---|---|---|
| $> d = 14$ | 21 | 21 | 21 | 21 | 21 | 20 | 21 | 15 | 21 |
| | User 10 | User 11 | User 12 | User 13 | User 14 | User 15 | User 16 | User 17 | User 18 |
| $> d = 14$ | 21 | 19 | 21 | 20 | 20 | 21 | 21 | 21 | 21 |
| | User 19 | User 20 | User 21 | User 22 | User 23 | User 24 | User 25 | | |
| $> d = 14$ | 20 | 20 | 21 | 21 | 21 | 19 | 21 | | |

Table 4. Number of comparisons with, at least, $d+1 = 22$ coincidences for $|S| = 192$

| | User 1 | User 2 | User 3 | User 4 | User 5 | User 6 | User 7 | User 8 | User 9 |
|---|---|---|---|---|---|---|---|---|---|
| $> d = 21$ | 21 | 20 | 21 | 20 | 21 | 19 | 19 | 14 | 18 |
| | User 10 | User 11 | User 12 | User 13 | User 14 | User 15 | User 16 | User 17 | User 18 |
| $> d = 21$ | 21 | 13 | 19 | 15 | 19 | 20 | 18 | 21 | 21 |
| | User 19 | User 20 | User 21 | User 22 | User 23 | User 24 | User 25 | | |
| $> d = 21$ | 16 | 15 | 20 | 21 | 18 | 17 | 20 | | |

**Table 5.** Number of comparisons with, at least, $d+1 = 29$ coincidences for $|S| = 256$

|  | User 1 | User 2 | User 3 | User 4 | User 5 | User 6 | User 7 | User 8 | User 9 |
|---|---|---|---|---|---|---|---|---|---|
| $> d = 28$ | 21 | 17 | 21 | 19 | 16 | 17 | 15 | 10 | 17 |
|  | User 10 | User 11 | User 12 | User 13 | User 14 | User 15 | User 16 | User 17 | User 18 |
| $> d = 28$ | 13 | 11 | 12 | 11 | 15 | 18 | 16 | 21 | 15 |
|  | User 19 | User 20 | User 21 | User 22 | User 23 | User 24 | User 25 |  |  |
| $> d = 28$ | 13 | 13 | 13 | 19 | 15 | 15 | 18 |  |  |

Considering that the total number of comparisons is $21 \cdot 25 = 525$, Table 6 shows the values of Genuine Acceptance Rate ($GAR$) and False Rejection Rate ($FRR = 1 - GAR$) for each value of $|S|$.

**Table 6.** Values of $GAR$ and $FRR$ for each value of $|S|$

| bitlength of $S$: $|S|$ | 64 | 128 | 192 | 256 |
|---|---|---|---|---|
| $GAR$ | 98.7% | 97.3% | 88.9% | 74.5% |
| $FRR$ | 1.3% | 2.7% | 11.1% | 25.5% |

## 3.2 Inter-user Variability: FAR

In this analysis, the templates of a given user are compared with all the templates of the rest of users. In this way a measure of the similarities among them is obtained.

This analysis is divided in two parts. In both parts of the analysis instead of using the 7 templates of each user, only one template is randomly chosen. In the first part, the templates considered are compared with the whole database (Templates vs. Database), and in the second part, the comparison is done only between the 25 chosen templates themselves (Templates vs. Templates). Then, two values for the False Acceptance Rate are obtained, $FAR_1$ and $FAR_2$, respectively.

**Analysis of Templates vs. Database**

In the first part of this analysis the 25 chosen templates are compared with the whole database formed by the 24 other users. The number of coincidences obtained for each value of bitlength are shown in Tables 7, 8, and 9 (for the value $|S| = 256$, there are only 3 coincidences, in the users 15, 18 and 22, so the table for this case is not shown).

Finally, taking into account all the values obtained, the False Acceptance Rate, $FAR_1$, for each bitlength, $|S|$, was computed, as it is shown in Table 10.

**Analysis of Templates vs. Templates**

In this part, the comparison is done only between the 25 chosen templates themselves, so in total there are 300 comparisons.

Table 11 shows the number of comparisons with, at least, $d+1$ coincidences and the corresponding value for the False Acceptance Rate, $FAR_2$.

**Table 7.** Number of comparisons with, at least, $d+1=8$ coincidences for $|S|=64$

|          | Tpl. 1  | Tpl. 2  | Tpl. 3  | Tpl. 4  | Tpl. 5  | Tpl. 6  | Tpl. 7  | Tpl. 8  | Tpl. 9  |
|----------|---------|---------|---------|---------|---------|---------|---------|---------|---------|
| $>d=7$   | 120     | 144     | 107     | 57      | 109     | 118     | 150     | 11      | 90      |
|          | Tpl. 10 | Tpl. 11 | Tpl. 12 | Tpl. 13 | Tpl. 14 | Tpl. 15 | Tpl. 16 | Tpl. 17 | Tpl. 18 |
| $>d=7$   | 102     | 88      | 110     | 134     | 115     | 143     | 159     | 146     | 136     |
|          | Tpl. 19 | Tpl. 20 | Tpl. 21 | Tpl. 22 | Tpl. 23 | Tpl. 24 | Tpl. 25 |         |         |
| $>d=7$   | 90      | 104     | 121     | 107     | 147     | 130     | 109     |         |         |

**Table 8.** Number of comparisons with, at least, $d+1=15$ coincidences for $|S|=128$

|          | Tpl. 1  | Tpl. 2  | Tpl. 3  | Tpl. 4  | Tpl. 5  | Tpl. 6  | Tpl. 7  | Tpl. 8  | Tpl. 9  |
|----------|---------|---------|---------|---------|---------|---------|---------|---------|---------|
| $>d=14$  | 18      | 44      | 41      | 3       | 11      | 17      | 42      | 0       | 5       |
|          | Tpl. 10 | Tpl. 11 | Tpl. 12 | Tpl. 13 | Tpl. 14 | Tpl. 15 | Tpl. 16 | Tpl. 17 | Tpl. 18 |
| $>d=14$  | 18      | 3       | 25      | 35      | 35      | 52      | 65      | 55      | 59      |
|          | Tpl. 19 | Tpl. 20 | Tpl. 21 | Tpl. 22 | Tpl. 23 | Tpl. 24 | Tpl. 25 |         |         |
| $>d=14$  | 13      | 16      | 10      | 37      | 37      | 33      | 31      |         |         |

**Table 9.** Number of comparisons with, at least, $d+1=22$ coincidences for $|S|=192$

|          | Tpl. 1  | Tpl. 2  | Tpl. 3  | Tpl. 4  | Tpl. 5  | Tpl. 6  | Tpl. 7  | Tpl. 8  | Tpl. 9  |
|----------|---------|---------|---------|---------|---------|---------|---------|---------|---------|
| $>d=21$  | 0       | 2       | 2       | 0       | 0       | 2       | 4       | 0       | 0       |
|          | Tpl. 10 | Tpl. 11 | Tpl. 12 | Tpl. 13 | Tpl. 14 | Tpl. 15 | Tpl. 16 | Tpl. 17 | Tpl. 18 |
| $>d=21$  | 1       | 0       | 1       | 2       | 2       | 6       | 7       | 4       | 8       |
|          | Tpl. 19 | Tpl. 20 | Tpl. 21 | Tpl. 22 | Tpl. 23 | Tpl. 24 | Tpl. 25 |         |         |
| $>d=21$  | 0       | 0       | 0       | 9       | 1       | 2       | 4       |         |         |

**Table 10.** Values of $FAR_1$ for each bitlength of $S$

| bitlength of $S$ | 64     | 128    | 192   | 256   |
|------------------|--------|--------|-------|-------|
| $FAR_1$          | 67.78% | 16.79% | 1.35% | 0.07% |

**Table 11.** Values of $FAR_2$ for each bitlength of $S$

| bitlength of $S$  | 64     | 128   | 192   | 256 |
|-------------------|--------|-------|-------|-----|
| $>d$ coincidences | 179    | 29    | 2     | 0   |
| $FAR_2$           | 59.67% | 9.67% | 0.67% | 0%  |

## 4  Conclusions and Future Work

In this work, an analysis of the security efficiency of a fuzzy extractor scheme for iris templates is presented. The main conclusions are the following:

1. Referring to the global efficiency of the scheme, it can be stated that the lower the bitlength of $S$ is, the easier to recognize a known user. Therefore the lower the percentage of False Rejection Rate is, which is a good improvement.

2. Nevertheless, at the same time, the lower the bitlength of $S$ is, the easier to recognize an attacker as a known user, as the values of $FAR$ show.
3. As the False Acceptance Rate is a security relevant measure while the False Rejection Rate is more a comfort criteria, a commitment between these two values has to be taken, but giving more importance always to the $FAR$.
4. Thus, from the four different values of $|S|$ analyzed, the best one in relation to the intra-user variability is $|S| = 256$ ($FAR_1 = 0.07\%$ and $FAR_2 = 0\%$); whereas the best value in relation to the intra-user variability is $|S| = 64$ ($FRR = 1.3\%$). So, it cannot be stated in a definitive way what value of $|S|$ is the best. That value will depend on the security requirements of the application where this scheme will be used. Anyway, the most balanced solution from the values of $FAR$ and $FRR$ could be $|S| = 192$.

From the previous conclusions, it would be of interest to improve the implementation of the scheme in order to perform the experiments faster and to use bigger values for $|S|$, for example, 384, 512, etc. Moreover, it is important to improve the extraction algorithms for iris templates in order to reduce the inter-variability and increase the intra-variability of users.

**Acknowledgment.** This work has been supported by Ministerio de Industria, Turismo y Comercio (Spain) in collaboration with Telefónica I+D (Project SEGUR@) with reference CENIT-2007 2004.

# References

1. Diez Laiz, E.: Master Thesis, Universidad Politécnica de Madrid (to appear)
2. Dodis, Y., Ostrovsky, R., Reyzin, L., Smith, A.: Fuzzy Extractors: How to Generate Strong Keys from Biometrics and Other Noisy Data. SIAM Journal Computing 38(1), 97–139 (2008)
3. Hernández Álvarez, F., Hernández Encinas, L., Sánchez Ávila, C.: Biometric Fuzzy Extractor Scheme for Iris Templates. In: Proc. of The 2009 World Congress in Computer Science, Computer Engineering, and Applied Computing, WORLDCOMP 2009 (to appear, 2009)
4. Juels, A., Sudan, M.: A fuzzy vault scheme. Designs, Codes and Cryptography 38(2), 237–257 (2006)
5. Juels, A., Wattenberg, M.: A fuzzy commitment scheme. In: Proc. of the 6th ACM conference on Computer and Communications Security, pp. 28–36 (1999)
6. Jain, A.K., Nandakumar, K., Nagar, A.: Biometric Template Security. Journal on Advances in Signal Processing 8(2), 17 (2008)
7. Lee, Y.-J., Bae, K., Lee, S.-J., Park, K.R., Kim, J.H.: Biometric key binding: Fuzzy vault based on iris images. In: Lee, S.-W., Li, S.Z. (eds.) ICB 2007. LNCS, vol. 4642, pp. 800–808. Springer, Heidelberg (2007)
8. Menezes, A., van Oorschot, P., Vanstone, S.: Handbook of Applied Cryptography. CRC Press, Boca Raton (1997)
9. Tong, V.V.T., Sibert, H., Lecœur, J., Girault, M.: Biometric fuzzy extractors made practical: A proposal based on fingerCodes. In: Lee, S.-W., Li, S.Z. (eds.) ICB 2007. LNCS, vol. 4642, pp. 604–613. Springer, Heidelberg (2007)
10. Uludag, U., Pankanti, S., Prabhakar, S., Jain, A.K.: Biometric Cryptosystems: Issues and Challenges. Proc. of the IEEE 92(6), 948–960 (2004)

# Behavioural Biometrics Hardware Based on Bioinformatics Matching

Slobodan Bojanić[1], Vukašin Pejović[1], Gabriel Caffarena[1], Vladimir Milovanović[2], Carlos Carreras[1], and Jelena Popović[2]

[1] Universidad Poltécnica de Madrid, ETSI Telecomunicación, Ciudad universitaria s/n, 28040 Madrid, Spain
{slobodan,vule,gabriel,carreras}@die.upm.es
[2] University of Belgrade, Faculty of electrical Engineering, Bul. Kralja Aleksandra 73, 11000 Belgrade, Serbia
{vlada,jelena}@el.etf.bg.rs

**Abstract.** In this work we realized special hardware for intrusion detection systems (IDS) based on behavioural biometrics and using bionformatics' Smith-Waterman algorithm. As far as we know there are no published hardware implementations of bioinformatics algorithms used for IDS. It is shown in the paper that the use of hardware can efficiently exploit the inherent parallelism of the algorithm and reach Gigabit data processing rates that are required for current communications. Each processing unit can be replicated many times on deployed Field Programmable Gate Array (FPGA) and depending on the capacity of the device, almost proportionally increase the throughput.

**Keywords:** behavioural biometrics, intrusion detection, pattern recognition, FPGA, bioinformatics, Smith-Waterman algorithm, dynamic programming.

## 1 Introduction

Pattern matching is the most widespread attack detection technique [6]. It generally means matching text or binary sequences against known elements often referred to as signatures. An attack usually possesses some kind of signature that identifies it and can be described as a Boolean relation called rule. The technique basically looks for one specific signature in the data flow. This signature often consists in one or more specific binary patterns found in a given file or network packet.

In this work we tackle the intrusion detection using the behavioural biometrics approach [3] based on bioinformatics' Smith-Waterman algorithm generally used for the DNA sequence alignment [8]. Since the algorithm can be significantly accelerated exploiting inherent parallelism, we deployed a hardware implementation to reach Gigabit rates that are required in contemporary data processing [9]. As far as we know there is no published work on the hardware deployment in IDS of the above mentioned algorithm.

The target technology, Field Programmable Gate Arrays (FPGA) is suitable for a broad range of applications as current FPGAs contain much more configurable logic blocks (CLBs) than their predecessors. Some researchers suggest that FPGAs have become highly competitive with microprocessors in both peak and sustained performance [10]. Besides high computing performance, the current FPGAs also provide large amounts of on-chip and off-chip memory bandwidth to I/O-bound applications.

Masquerade is a security attack in which an intruder assumes the identity of a legitimate user. It is shown that using a bioinformatics algorithm as behavioural biometrics tool provides the best rates in terms of false positive and false negative alarms in the detection of the attack [4]. The Smith-Waterman bioinformatics algorithm was originally applied in gene alignment but efficient hardware implementations widen its use to other fields [1], [2].

The rest of the paper is organized as follows. Section 2 gives a background on applying the behavioural biometrics to detect masquerading attacks. Section 3 explains the Smith-Waterman algorithm. In Section 4 its application to intrusion detection is presented. Section 5 describes the hardware implementation and its benefits in comparison to software implementation. Section 6 presents the corresponding results. Finally the conclusions are drawn in Section 7.

## 2 Behavioural Bioinformatics Detection of Masquerading Attack

Masquerade attacks when an attacker assumes the identity of a legitimate user often occur when an intruder obtains a legitimate user's password or when a user leaves their workstation unattended without any sort of locking mechanism in place. It is difficult to detect because the attacker appears to be a normal user with valid authority and privileges. The damage that can be done like stolen/destroyed documents, data, e-mail, makes them a very serious threat to computer and network infrastructure.

To detect a masquerader, behavioural biometrics makes a contrast between the real legitimate user and the intruder. Each user has its own specific behavior that makes a so called user signature. Comparing it with the behavior of logged user, the matching differentiates a legal use and an intrusion. The user signature usually contains a sequence of commands, but it could also contain a user style of typing on a keyboard or specific mouse movements, biometric features.

Since the user behavior can change over time, the signatures should be updated frequently. In a small period a user can react in a different manner and in long period can change the behavior fundamentally. Physiological biometrics like fingerprint or eye recognition is more reliable and unvarying over time but in remote systems it can be useless. Combining physiological biometrics provides better results but is complex to achieve in real time on general purpose computers.

Several detecting algorithms have been developed with more or less success in finding anomalies. An intruder may happen to have similar behaviour and/or the user's change in activity is not captured strongly in the signature. In [4] various masquerade detection techniques like Bayesian, Markov model etc. were analyzed in terms of missed attacks as false negatives and false alarms as false positives. The best results exhibited the bioinformatics approach that uses slight modification of the Smith-Waterman algorithm, originally published for the purpose of gene alignment.

## 3  Bioinformatics Appliance to Intrusion Detection

Sequence alignment is already a well-studied tool used to quantify and visualize similarity between two or more sequences. It is originally developed for the comparison of genetic material, such as DNA, but it can also be used as local alignment algorithm for other sequences, strings or in this case user signatures.

The Smith-Waterman (SW) algorithm is a database search algorithm developed for use in bioinformatics [8], and based on an earlier model named Needelman-Wunsch [5]. The SW algorithm applies a dynamic programming technique which takes alignments of any length, at any location, in any of two input sequences, and determines whether an optimal alignment can be found. Based on these calculations, scores or weights are assigned to each character-to-character comparison (positive score for exact matches and substitutions, negative score for insertions and deletions) and the highest scoring alignment is reported.

Because of its complexity, many heuristic methods like BLAST and FASTA were developed. But the original SW algorithm is superior to them as it searches a larger field, that makes it a more sensitive technique [2]. However, individual pair-wise comparisons between letters slow down the process significantly. Instead of looking at an entire sequence at once, the SW algorithm compares multi-length segments, looking for whichever segment maximizes the scoring measure. The algorithm itself is recursive in nature:

$$H_{i,j} = \max \begin{cases} H_{i-1,j-1} + s(a_i, b_j), \\ \max_{k \geq 1} \{H_{i-k,j}, W_k\} \\ \max_{l \geq 1} \{H_{i,j-l}, W_l\} \end{cases}$$

where $a_i$ and $b_j$ are units of sequences that are being compared on positions $i$ and $j$, respectively, $H_{i,j}$ is matrix value on position with coordinates $i, j$, respectively, and $W_k$ is weight for deletion of $k$ sequence units [8].

Instead of comparing nucleotides, the concept can be used to align the sequences of commands: user signature corresponds to one sequence and the test block corresponds to another. If the algorithm is able to align some subsequences extracted from those two arrays, scoring them adequately, and if the score is high enough it could mean that the legitimate user is logged on. If the alignment score is not sufficient, the potential intruder is logged and the session should be interrupted.

The original algorithm formula can be modified as the following formula:

$$M_{ij} = \max\{(M_{i-1,j-1} + s_{ij}), (M_{i,j-1} + g_v), (M_{i-1,j} + g_h), 0\}$$

where $M_{ij}$ is a weight matrix element in $i$-th row and j-th column., $s$ some positive reward for match interpreting similarity, $g_h$ and $g_v$ are usually negative horizontal and vertical gap penalties, respectively.

In this case it is possible to apply the following principle. We want to be able to insert gaps into the tested block to simulate the insertion of commands between characteristic groups of commands in the user's signature. This requires to provide a slightly lesser penalty for gaps in the tested block. Matches should positively influence the score of an alignment, and should be chosen so that matches are preferred to gaps. Using the above criteria, we chose scores of +1 for a match between two aligned commands, -2 for a gap placed in the tested block, -3 for a gap placed in the user's signature, and 0 for a mismatch between aligned commands.

Mismatches are kept at a constant score of 0, as a blanket reward or penalty for any mismatch would unfairly favor certain alignments, and would not disallow concept drift [3]. The scoring system rewards the alignment of commands in the user segment but does not necessarily penalize the misalignment of large portions of the signature that do not necessarily align with a segment of the user's commands.

The scores are taken from [3], but any other scoring can be applied in our implementation without any degradation in performance. As the goal is to align characteristic groups of commands in a tested block with similar groups in the user's signature the idea is to heavily penalize any gaps within the signature itself, because we do not want commands in the tested block to be aligned with gaps in the user's signature. Semi-global Smith-Waterman comparison algorithm [3], is very good starting point as it has hit-rate of 75.8%, while false positive being 7.7%.

Although having the best detection results in the masquerade intrusion, the sequentially implemented algorithm is quite slow. The complexity of the algorithm is $O(mn)$ where $m$ and $n$ are the lengths of the user signature and test block respectively. This is not a problem if the sequences are relatively short, but if the biometric approach is added the sequences can even reach lengths up to a million elements. The fastest general purpose computer can not accomplish this task in real time. The algorithm contains no multiplication that can slow down the calculations but the amount of data that needs to be processed is enormous.

## 4 Software and Hardware Implementation

The main drawback of the Smith-Waterman algorithm is its slowness in sequential implementation. Thus many heuristic methods have been developed e.g. FASTA and BLAST but they produce too many errors in detecting the masquerading intrusions.

To illustrate a software implemented Smith-Waterman algorithm we programmed one in ANSI C, where integers are used instead of the originally used floating point values. This way it uses twice less memory than the original floating point algorithm. On AMD AthlonXP 2GHz clock frequency and 256KB cache memory based PC processor with 256MB of RAM, the average execution time needed to calculate one score-matrix element was $13.3 \mu s$ and the worst case execution time of the same action was $16.4 \mu s$. In order to generate one score-matrix element two characters i.e. 2-byte info needs to be introduced that means that throughput of software system would be approximately 1.2Mbps.

The only way to reduce the complexity is to introduce the parallelism. It cannot be applied on general purpose machines where the processor has only one arithmetic-logic unit. But hardware devices like FPGAs have capabilities to bring the necessary parallel calculations and reduce the complexity, and set free other hardware parts that were used in processing the user signature and the test block.

The computation of a weight matrix can be implemented in several different ways. The first way can be like in sequential machines. For calculating the current element $M_{ij}$, we need to have the diagonal element $M_{i-1,j-1}$, the upper element $M_{i-1,j}$ and finally the left element, $M_{i,j-1}$. Using these three as well as reward/penalty constants we can produce the result. We can create a hardware cell, a simple processing element, and optimize it for weight matrix calculation. One such processing element is shown in Fig. 1.

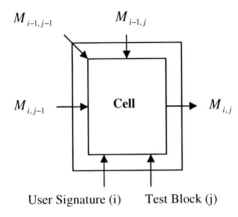

**Fig. 1.** Bioinformatics processing element

The computation of a weight matrix can be implemented in several different ways. The first way can be like in sequential machines. For calculating the current element $M_{ij}$, we need to have the diagonal element $M_{i-1,j-1}$, the upper element $M_{i-1,j}$ and finally the left element, $M_{i,j-1}$. Using these three as well as reward/penalty constants we can produce the result. We can create a hardware cell, a simple processing element, and optimize it for weight matrix calculation. One such processing element is shown in Fig. 1.

But, it is still possible to reduce the time complexity from quadratic to linear without letting area to increase that much with an increase of the sequence length for a matrix size *m×n* using a systolic array of at least min{*m,n*} processing elements. Although the user signature is by the rule almost always longer than the test block, it is better to assign number of elements that correspond to a signature's length for the reasons that will be later explained. This is a feasible solution having in mind the area that this hardware occupies. The same processing element calculates distances in the column in which it is placed (Fig. 2).

Therefore the whole matrix will be calculated in *m+n-1* clock cycles with a linear complexity of *O(m+n)*. When the length of sequences rise prevails linear time complexity and linear area occupation determine the method of choice.

There is no need to load both, the user signature and the test block, in the device concurrently. While the user is logging on, after the correct password is imported, the

|   | 0 | 1 | 2 | 3 |
|---|---|---|---|---|
| 1 |   | $M_{1,1}$ $PE_{1(T1)}$ | $M_{2,1}$ $PE_{2(T2)}$ | $M_{3,1}$ $PE_{3(T3)}$ |
| 2 |   | $M_{1,2}$ $PE_{1(T2)}$ | $M_{2,2}$ $PE_{2(T3)}$ |   |
| 3 |   | $M_{1,3}$ $PE_{1(T3)}$ |   |   |

**Fig. 2.** Parallelized matrix calculation

user signature is loaded to the FPGA as a part of the loading process that occurs until system is ready to work. This eliminates demands for storing user signatures in the FPGA whose memory is often very small and can be used for other purposes, but in some other memory like flash or hard disk drive. After the session starts, all the operations of interest are monitored and rewritten creating the test block for comparison with user signature. The test block is compared with the user signature as the commands are inserted by the user and/or the specific movements of mouse and keyboard occur giving that some sort of pipelining in the system, yielding very fast detection, just after the last needed command for decision whether or not the legitimate user is logged to the system.

## 5 Results and Discussion

We implemented a general Smith-Waterman algorithm with the score that can be any integer of fixed point decimal value. We used one of the most popular FPGA families, Xilinx Virtex 4LX200. In structures like FPGAs the best way to gain speed is to create highly optimized regular cells and then simply replicate them. We assumed that there are 256 used commands, which in operating systems like, UNIX and Linux are more than enough. We also assumed that the characteristic mouse movements and keyboard typing styles can be represented with eight bits each, although their expansion would not dramatically decrease performance.

The circuit was implemented using Xilinx ISE 7.1i tool. The systolic cells were first implemented in VHDL and then optimized by manually placing their inner components and creating relational-placed macros (RPM). The RPMs allowed an efficient use of the FPGA and higher processing speed. Both area and speed were improved by an approximate factor of a 100%. The area was reduced exploiting the regularity of the systolic array through the creation of cells that were able to fit more precisely within the overall area. The speed was doubled reaching clock frequencies of 200 Mhz due to the reduction of routing lines within each systolic cell and among cells.

In a Xilinx Virtex 4 one cell takes 100 logic cells, what is 50 slices (or 25 CLBs). This means that more than 2000 processing elements can be implemented on this chip. On the 8-bit wide bus, the equivalent throughput is 250MB/s or 2Gb/s that is a number of orders more than for sequential implementation. Self equipped devices based on microprocessor and FPGA combining the network support gives the possibility to implement this design independently from personal computer, and monitor the network independently and thus free the server totally from this activity and, what is also important, to be placed somewhere else.

The results are obtained from the simulation with partially manually generated user signatures as the input. Exactly, we have written a script to encode and precisely enumerate the set of the commands available on the host system. The commands are thus ordered by their alphabetic order and an ordinal number is assigned to each and every one of them. The encoding script was written to enumerate the 256 mostly used commands. This has led to the matrix construction by the comparison of the 8-bits wide command elements. The members of our investigation group have been asked to track and provide their own profiles, which they have in most of the cases done manually, and these entries were then processed by the enumeration script and used as the simulation data. This process, however, has to be completely automated before being used on a real system.

The described simulation input data was fed to the comparison engine. After the construction of the matrix the decision logic analyses the obtained values. If the maximum value encountered in the matrix was higher than 80% of the length of the user signature, the logic permits the further interaction of the user and the host. Similarly, the other scenarios can include different decisions based on the maximum similarity value recorded in the matrix. In continuation we need to study the options for the design of the loop closure mechanism to be used for the dynamic determination of the threshold value, aforementioned and fixed 80%.

Within the network environment, each workstation would need to have its own masquerade detecting device. To exploit the throughput capabilities of the proposed schema the device will have to be attached to the PCI Express bus with at least 8 lanes and resulting throughput of 8x250Mbps=2Gbps. That would than allow an automated mechanism, sketched in the paragraph above to be used as continuous feeding engine of the detection system.

The server host, whatsoever, requires a different approach. Besides processing the data the same way the workstations do, locally it needs to consider the remote user actions and profiles derived from those. The monitoring of the secure sessions and encoding of the information would be unavoidable for such a case. Furthermore, the issues such as a variable latency that network provides and the influence of this phenomena to the user profiles generation process, and later comparison and decision processes needs to be thoroughly studied before having a server based system. Yet, from the server perspective we note the importance of throughput performance the proposed "de-masquerading" schema is capable of delivering. We expect to undertake both the simulation based and practical tests of the system and obtain an objective efficiency measure.

## 6 Conclusion

We designed special intrusion detection hardware based on behavioural biometrics. The pattern recognition is carried out using the bioinformatics Smith-Waterman algorithm that exhibits high performances in detecting masquerading attack. The slowness of the algorithm is significantly reduced due to parallel hardware implementation.

We designed hardware architecture that can deploy any sequence alignment scoring in contrast to bioinformatics case where only fixed values are applied and optimized. The design was deployed in Virtex 4 FPGA and, as far as we know, there is no published hardware IDS implementations that uses this bioinformatics approach.

Each hardware unit reaches Gigabit processing rates thus satisfying contemporary data flow requirements and for several orders overcoming sequential software solutions. Due to optimized area design, the units can be massively replicated on FPGA devices and achieve a significant throughput increase.

**Acknowledgment.** The work was supported by Ministry of Science of Spain through the project TEC2006-13067-C03-03 and EC FP6 project IST 027095.

## References

1. Bojanić, S., Caffarena, G., Petrović, S., Nieto-Taladriz, O.: FPGA for pseudorandom generator cryptanalysis. Microprocessors and Microsystems 30(2), 63–71 (2006)
2. Caffarena, G., Pedreira, C., Carreras, C., Bojanić, S., Nieto-Taladriz, O.: FPGA acceleration for DNA sequence alignment. Journal of Circuits, Systems and Computers 16(2), 245–266 (2007)
3. Coull, S., Branch, J., Szymanski, B., Breimer, E.: Intrusion Detection: A Bioinformatics Approach. In: Proc. 19th Annual Computer Security Applications Conf, p. 24 (2003)
4. DuMouchel, W., Ju, W.H., Karr, A.F., Schonlau, M., Theusen, M., Vardi, Y.: Computer Intrusion: Detectiong Masquerades. Statistical Science 16(1), 58–74 (2001)
5. Needleman, S.B., Wunsch, C.D.: A general method applicable to the search for similarities in the amino acid sequence of two proteins. Journal of molecular biology 48, 443–453 (1970)
6. Sai Krishna, V: Intrusion Detection Techniques: Pattern Matching and Protocol Analysis, Security Essentials (GSEC) Version 1.4b, April 26 (2003), http://www.giac.org/
7. Sinclair, C., Pierce, L., Matzner, S.: An Application of Machine Learning to Network Intrusion Detection. In: Proc. 1999 Anual Computer Security Applications Conf. (ACSAC), Phoenix, USA, December 1999, pp. 371–377 (1999)
8. Smith, T.F., Waterman, M.S.: Identification of common molecular subsequences. Journal of molecular biology 147, 195–197 (1981)
9. Tripp, G.: An intrusion detection system for gigabit networks – architecture and an example system. Technical Report 7-04, Computing Laboratory, University of Kent (April 2004)
10. Underwood, K.D., Hemmert, K.S.: Closing the gap: CPU and FPGA trends in sustainable floating-point BLAS performance. In: Proc. 2004 IEEE Symposium on field-programmable custom computing machines (FCCM 2004), USA (2004)

# Robust Real-Time Face Tracking Using an Active Camera

Paramveer S. Dhillon

CIS Department, University of Pennsylvania, Philadelphia, PA 19104, U.S.A

**Abstract.** This paper addresses the problem of facial feature detection and tracking in real-time using a single active camera. The variable parameters of the camera (i.e. pan, tilt and zoom) are changed adaptively to track the face of the agent in successive frames and detect the facial features which may be used for facial expression analysis for surveillance or mesh generation for animation purposes, at a later stage. Our tracking procedure assumes planar motion of the face. It also detects invalid feature points i.e. those feature points which do not correspond to actual facial features, but are outliers. They are subsequently abandoned by our procedure in order to extract 'high level' information from the face for facial mesh generation or emotion recognition which might be helpful for Video Surveillance purposes. The only limitation on the performance of the procedure is imposed by the maximum pan/tilt range of the camera.

## 1 Introduction

Facial feature detection and tracking is being extensively studied in computer vision for a variety of applications which vary from surveillance to facial expression analysis. The increased interest in this field is evident by the quantity of literature written in this field and the IEEE and IEE workshops and conferences on human motion analysis and special journal issues that focus solely on gesture recognition, video surveillance and human motion analysis.

Normally, the aforementioned tasks of detection and tracking are accomplished by static cameras or multiple active cameras working in a cooperative manner [2]. The predicament involved in tracking with a static camera is that the motion causes the image to be blurred in the area of motion as shown in Fig. 1(a), (b) and sometimes the object of interest may be improperly small or large, and in some cases it may disappear from the image i.e the field of view of camera, altogether. All these problems are inherently associated with the 'passiveness' of the static camera and cannot be rectified. So, the alternative is to use active cameras which are immune to all the above mentioned problems. One approach is to use multiple active cameras. But, the problem with multiple active cameras is that, firstly, the system will be computationally expensive and may prove to be a big resource hog and secondly, it is not fault tolerant. So, the best approach is to use a single active camera in simple and moderately complex scenes. This paper addresses the issue of feature detection and tracking of human faces in image space, using a single active camera. The pan-tilt-zoom

**Fig. 1.** Left: Image of a moving object by static camera. Right: Image of a moving object by active camera.

parameters of the camera are varied adaptively depending on the velocity of agent, so that the agent i.e. the human face remains in camera's field of view throughout the tracking range of camera i.e within the limits of maximum pan-tilt of the camera. After initialization of the tracking module, the feature tracker detects and tracks good features [3], and subsequently 'culls' or rejects spurious features. Hence, we are able to extract 'high level' information from these facial features, for mesh generation or emotion recognition etc. The remaining paper is organized as follows: Section 2 deals with previous work done in this field, Section 3 deals with notations used in this paper. Section 4 discusses the main procedure of the algorithm. Section 5 deals with experimental results and finally, Section 6 concludes the work by providing a summary.

## 2 Related Work

The problem of facial feature detection and tracking using an active camera has been studied by many researchers over the last few years. The concept of tracking motion based on optical flow method using an active camera has been studied by Murray et al. [4]. They have also elucidated the advantages of an active camera over a static camera in motion detection. The controlling of zoom of the active camera for activity recognition has been studied by Smith et al. [5].They use labelling and trajectory constraints to detect head, hands etc. which are later used for activity recognition. Lately systems of multiple active cameras have also gained attention of researchers. A multiple camera system having co-operation amongst the sensors (cameras) is studied in [2]. In this paper they advance the concept of dynamic memory and AVA (Active vision agent i.e a network connected computer with an active camera). Another piece of related work is [1], which provides a system which is robust to scale change or partial occlusion of face. The authors of [1] match the candidate face with a face model using Bhattacharya's coefficient as a measure of similarity and use mean shift analysis for tracking.

## 3 Notations Used

In this paper we use certain terms and notations, which are explained here and visualized in Fig. (2). Firstly, the *agent* is the person i.e. the face that is currently

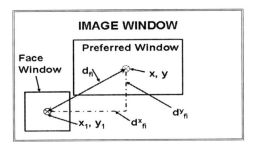

**Fig. 2.** Figure showing the Notations Used

in the focus of the active camera. At t = 0 i.e the position of zero pan/tilt, the position of the camera is called the *neutral position*, $s_{tilt}$ and $s_{pan}$ are the steps in which the tilt and the pan of the camera are changed, from the neutral position and $v_1$ is the velocity of the agent's centroid in real-world. Since we analyze the motion of the agent in the frame of reference of the camera, so we define $v_{is}$ as the velocity of the agent in image space. Throughout the course of tracking we assume planar motion of the agent.

Next, we define some terms specific to the active camera. A measure of the camera's capability is $d_{PTZ}$ which is the maximum real-world distance which the camera can cover. Similarly, we have $\triangle_{PTZ}$ which is the maximum angular distance that the camera can move $[-\pi, +\pi$ radians] for pan and $[-\frac{\pi}{6}, +\frac{\pi}{6}]$ for tilt, in our case. Another important term is $t_{total}$ which denotes the time taken by the agent to move beyond the panning/tilting range of the camera, starting at t=0 from neutral position of the camera. $fps$ which is the number of frames per second captured by the active camera. Lastly, we describe the terms related to the analysis in image space. So, we have a window showing the LIVE video captured from the camera, this window is called the *image window*. The part of the image window containing the face of the agent i.e the region of interest, is called the *face window*, which is bounded by a rectangle. $d_{face}$ is the euclidean distance between the centres of face window in two consecutive frames captured by the camera. Next, we have the *preferred window* which is a window formed by the region (100 × 100) pixels with its centre coinciding with that of image window. It is also a part of the image window. Now, we explain some notations related to these windows, $(x, y)$ and $(x_1, y_1)$ are the cartesian co-ordinates of the centre of the image window and face window respectively. Related to these is $d_{fi}$ which is the euclidian distance between the centres of image window and face window. Mathematically it is given by $\sqrt{((x - x_1)^2 + (y - y_1)^2)}$ and $d_{fi}^x$ and $d_{fi}^y$ are the X and Y components of $d_{fi}$ respectively.

## 4 Main Procedure

The procedure has been divided into two separate modules *Active Camera Module* and *Detection and Tracking Module*, which supplement each other and ultimately unite to form the complete system.

## 4.1 Active Camera Module

The biggest advantage of an Active Camera is that its external parameters can be modified and hence its optical behaviour can be changed. In this manner, we can change the horizontal and vertical orientation i.e pan and tilt respectively, of the camera through software, and make it to focus on the object or region of interest. In our case, we need to detect and track facial features using a single active camera,so the first step is to detect the face. The treatment of face detection is given in next subsection. After the face has been detected, it is chosen as the region of interest by inscribing it in a rectangle, which we call as the face window and it is zoomed to a predefined value to achieve uniform size in all successive frames, so that we get a better perspective and can easily compare a frame with earlier one to compute the displacement of the centre of face window, as will be described later.

Now, the temptation is to bring the face window in the central region of the image window. This central region is called preferred window and its expanse is (100 × 100) pixels from the centre of image window (x,y). The tendency to do this is that if the agent is in preferred window then it has relatively lesser chance of going out of field of view of camera, without allowing the active camera to modify its pan/tilt to track it. Or in other words we can say that if agent is not in Preferred Window then there is a high probability that the agent runs out from the field of view of camera. There can be many reasons for the running out of the agent like bending of the agent to tie his/her shoe lace or a casual increase in the velocity of agent, i.e a sudden discontinuity in usual state of affairs will make an agent untrackable in frame '(n+1)' although s/he was trackable in frame 'n'.

If $P(x_i)$ is the probability of the agent being tracked in a given frame $n = n_1$ then:

$$P_n(x_i) = f(\frac{1}{d_{fi}}) \tag{1}$$

In order to bring the agent in the preferred window, we proceed by calculating $d_{fi}$ and its x and y components i.e $d^x_{fi}$ and $d^y_{fi}$ respectively, now we have

$$\frac{\partial(d^x_{fi}, d^y_{fi})}{\partial t} \leq T, \tag{2}$$

i.e if the rate of change of $d^x_{fi}$ and $d^y_{fi}$ is less than a threshold T, then the face window is in 'safe' region which is in fact the preferred window, so the value of T is chosen accordingly, but if the value of the derivative is greater than the threshold T, then we see which component changes faster $d^x_{fi}$ or $d^y_{fi}$ and decide whether pan or tilt need to be modified earlier, by considering the fact that the component which changes at a faster rate needs to be modified earlier because it will 'run out' of the camera's field of view earlier. So, we summarize the above theory by following equations:

$$s_{tilt} = f_{tilt}(d^y_{fi}) \tag{3}$$

$$s_{pan} = f_{pan}(d^x_{fi}) \tag{4}$$

Hence, we can decide which parameter i.e pan or tilt to modify earlier, based on the above hypothesis.

Once the agent is in preferred window the 'run out' problem ceases to be a serious problem. Therefore the tracking module is initialized as described in next subsection. Let us assume that tracking is initialized at time $t = t_1$, then by definition

$$d_{face} = [(x_1, y_1)|_{(n+1)} - (x_1, y_1)|_n] \quad (5)$$

Now, based on above equation we develop a hypothesis to adaptively modify the camera parameters. Since, the above equation gives a measure of motion of the agent in the image space so we can infer approximately the real-world velocity of the agent which enables us to modify the camera parameters i.e pan/tilt adaptively to track the agent effectively. Now the camera's pan/tilt parameters can be changed adaptively in a linear manner, depending on the velocity of agent $v_1$ and $d_{PTZ}$.

A measure of the velocity of the agent in the image space may be provided by the product of $d_{face}$ i.e the distance moved by the centre of face window in two successive frames, and the number of frames per second captured by the camera called $fps$. Now, under the assumption of planar motion of the agent's face, the velocity of the agent in image space is proportional to the actual velocity of agent in real-world, where the constant of proportionality will depend on the internal parameters of the camera.

We can put all these things mathematically as below:

$$v_{is} = d_{face} * fps \quad (6)$$

$$v_1 \propto v_{is} \quad (7)$$

Reverting back to the problem of changing camera parameters pan/tilt adaptively, we have an estimate of $v_1$ as explained in Eq.(7) and also $\triangle_{PTZ}$, the maximum angular distance that the camera can move, is known for a given camera, from which we can find $d_{PTZ}$ by proper calibration. So, now we have two constraints on the pan/tilt of the camera and they are $v_1$ and $d_{PTZ}$ i.e in time $\frac{d_{PTZ}}{v_1}$ the camera must reach the extreme angular position, provided that initially it is in neutral position i.e position of zero pan/tilt. Hence we can

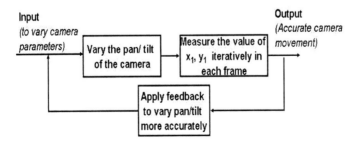

**Fig. 3.** Flowchart showing 'Feedback'

model the function used to vary the steps in which the pan/tilt should be varied. But, there is an assumption in the hypothesis that $v_1$ is constant, which usually is not the case, so in order to make the procedure more robust we introduce feedback by measuring the value of $d_{face}$ and hence $v_1$ at fixed timestamps like $[t_1, t_2........t_n]$ and then modifying the function which is used to model the step size of pan/tilt variation accordingly. This feedback makes the system relatively insensitive to variations in $v_1$. The flowchart showing the feedback mechanism is shown in Fig. (3).

The feedback is introduced in the system by calculating the velocity of agent in image space $\frac{\partial [x_1, y_1]}{\partial t}$. The derivative can be quantified by measuring $(x_1, y_1)$ in each frame i.e after a time of $\frac{1}{fps}$ seconds, in real time it can be achieved by programming a *multithreaded* application, in which a separate thread takes care of finding $[x_1, y_1]$ in each frame as described above.

## 4.2 Detection and Tracking Module

In this subsection, we analyze the procedure implemented to detect face and its subsequent tracking.

The face detector used is a modified version of the one described in [6], which uses the cascades of boosted classifiers of Haar-like features. We first localize the face based on skin colour and then run the Viola -Jones detector in that localized region. The ROC curve of our face detector is shown in Fig. 4. The results are comparable to the ones given in [6].

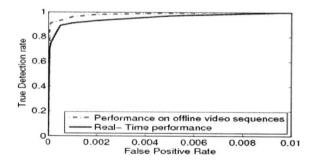

**Fig. 4.** ROC Curve

After the face is detected, the good facial features are tracked as described in [3]. As a modification to the procedure we make it more robust by 'culling' (rejecting) the feature points which have not been updated for quite a long time, because there is a probability that they have arisen by some noise and are not the good points that we were looking for. So, we define a parameter $t_c$ called time parameter or culling parameter, which decides the time constraints for culling of unupdated points. If the application is accuracy sensitive then $t_c$ should be more, otherwise if the application is speed sensitive then $t_c$ may be chosen to be less. Since in our case, accuracy was the major goal so we used a high value of $t_c$.

**Fig. 5.** Top Down Approach: From Polygonal Mesh Generation to Face Detection

The approach followed by us is the bottom-up approach, in which we first detect the face then mark robust feature points and ultimately do the mesh generation. But, due to the presence of noisy feature points and a large value of $t_c$, we construct the samples using top-down approach as shown in Fig. (5). These results are the results of our procedure on offline video sequences (mainly TV shows etc.), that we used to test our algorithm before doing its real- time implementation.

## 5 Results

The active camera used in our experiments was a PTZ (Pan/Tilt/Zoom) camera having a pan range of $[ -\pi , +\pi ]$ and tilt range of $[ -\frac{\pi}{6} , +\frac{\pi}{6}]$. The camera supports Motion JPEG and MPEG-4 formats with a frame rate of 15 fps. So, initially the face detector is initialized and its output is shown in Fig. 6 (a), (c). The detected face is shown with a bounding rectangle. Next, the detected agent is brought into preferred window. It is followed by initializing the tracking module which marks the strong features on the agent. These marked features are shown in Fig. 6 (b), (d), later on these features are tracked by a separate 'thread' by measuring their x-y position in image space, which in turn provides a measure for changing the pan/tilt of the camera depending on the velocity of the agent, as described earlier. As can be seen in Fig. 6 (b), (d) that some points are outside the face and moreover they are not updated in succeeding frames, so they are 'culled' in later frames, but since our application is accuracy sensitive we used quite a high value of $t_c$, hence the 'culling' occurs very slowly.

**Fig. 6.** Experimental Results: Left to Right (a - d): (a), (c) Face detected by the detector. (b), (d) Robust features of the face.

## 6 Conclusion and Future Work

In this paper we discussed an approach for detecting and tracking human facial expressions using a monocular active camera. The results shown before agree with the hypothesis to a high degree and the quality of results provides a good indication as to how this approach can be used for facial mesh generation for animation purposes or analyzing the expressions, emotions and gestures for video surveillance purposes. The problem of noisy, invalid feature points is there but their proper 'culling' was done. In our implementation the culling was quite slow due to the reasons mentioned earlier.

As a follow-up to this we propose to do high-level operations on the extracted facial images, such as those mentioned in the preceeding paragraph. Secondly, it would be interesting to integrate this approach with multiple levels of zooming for activity analysis as explained in [5].

# References

1. Comaniciu, D., Ramesh, V.: Robust Detection and Tracking of Human Faces with an Active Camera. In: VS 2000: Proceedings of the Third IEEE International Workshop on Visual Surveillance (VS'2000), vol. 02, p. 11 (2000)
2. Ukita, N., Matsuyama, T.: Real-time cooperative multi-target tracking by communicating active vision agents. icpr, vol. 02, pp. 2001-4 (2002)
3. Shi, J., Tomasi, C.: Good features to track. In: IEEE Conference on Computer Vision and Pattern Recognition (CVPR 1994) (June 1994)
4. Murray, D., Basu, A.: Motion tracking with an active camera. IEEE Transactions on Pattern Analysis and Machine Intelligence 16(5), 449–459 (1994)
5. Smith, P., Shah, M., da Vitoria Lobo, N.: Integrating multiple levels of zoom to enable activity analysis. Computer Vision and Image Understanding 103(1), 33–51 (2006)
6. Viola, P., Jones, M.: Robust real-time object detection. International Journal of Computer Vision (2001)

# An Approach to Centralized Control Systems Based on Cellular Automata

Rosaura Palma-Orozco[1], Gisela Palma-Orozco[1], José de Jesús Medel-Juárez[2], and José Alfredo Jiménez-Benítez[2]

[1] Escuela Superior de Cómputo del IPN, Juan de Dios Bátiz s/n, Zacatenco, 07738, México D. F.
[2] Centro de Investigación de Ciencia Aplicada y Tecnología Avanzada del IPN, Legaria 694, Irrigación, 11500, México D. F.

**Abstract.** Computational Intelligence (CI) embraces techniques that use Fractals and Chaos Theory, Artificial immune systems, Wavelets, etc. CI combines elements of learning, adaptation and evolution to create programs that are, in some sense, intelligent. Cellular Automata is an element of Fractals and Chaos Theory that is adaptive and evolutionary, so we are interested in using this approach for solving problems of centralized control. The main objective of the paper is to present the cellular automata as a model of centralized control of dynamic systems. Any dynamic system is subjected to conditions of internal and external behavior that modify its operation and control. This implies that the system can be observable and controllable. The authors take on the task of analysis of an example control traffic system. For the approach, is proposed a one-dimensional cellular automaton in an open-loop scheme.

## 1 Introduction

Computational Intelligence embraces techniques that use Fractals and Chaos Theory, Artificial immune systems, Wavelets, etc. It combines elements of learning, adaptation and evolution to create programs that are, in some sense, intelligent. Cellular Automata is an element of Fractals and Chaos Theory that is adaptive and evolutionary, so we are interested in using this approach for solving problems of centralized control. On the basis of the current trends for solving complex technical problems, a different approach of centralized control system is proposed. One of the essential problems in the presented conception is the answer to the question:

*It is possible to define a centralized control system using cellular automata?*

To answer the question formulated this way, an example of control traffic system, has been considered. Traffic phenomena attract much attention of physicists in recent years.It shows a complex phase transition from free to congested state, and many theoretical models have been proposed so far [6]. Among them we will focus on deterministic cellular automaton (CA) models. CA models are simple, flexible, and suitable for computer simulations of discrete phenomena.

The rule 184 CA[7] has been widely used as a prototype of deterministic model of traffic flow. In the model, lane is single and cars can move by one site at most every time step. Its several variations have been proposed recently.

## 2 Cellular Automaton, Rule 184 and Traffic Flow

A *cellular automaton* (plural: cellular automata) is a discrete model studied in computability theory, mathematics and systems modeling. It consists of a regular grid of cells, each in one of a finite number of states, "On" and "Off" for example. The grid can be in any finite number of dimensions. For each cell, a set of cells called its neighborhood (usually including the cell itself) is defined relative to the specified cell. For example, the neighborhood of a cell might be defined as the set of cells a distance of 2 or less from the cell. An initial state (time t=0) is selected by assigning a state for each cell. A new generation is created (advancing t by 1), by making the new state of each cell a value according to a rule that depends on the current state of the cell and the states of the cells in its neighborhood. For example, the rule might be that the cell is "On" in the next generation if exactly two of the cells in the neighborhood are "On" in the current generation, otherwise the cell is "Off" in the next generation. The rule for updating must be the same for each cell and does not change over time. Each generation the rules are applied to the whole grid simultaneously. [1][2]

Rule 184 is a one-dimensional binary cellular automaton rule, notable for solving the majority problem as well as for its ability to simultaneously describe several, seemingly quite different, particle systems. The earliest research on rule 184 seems to be the papers by Li (1987) and Krug and Spohn (1988). In particular, Krug and Spohn already describe all three types of particle system modeled by rule 184.

Rule 184 can be used as a simple model for traffic flow in a single lane of a highway, and forms the basis for many cellular automaton models of traffic flow with greater sophistication. In this model, particles (representing vehicles) move in a single direction, stopping and starting depending on the cars in front of them. The number of particles remains unchanged throughout the simulation. Because of this application, rule 184 is sometimes called the *traffic rule*. [8]

A state of the rule 184 automaton consists of a one-dimensional array of cells, each containing a binary value (0 or 1). In each step of its evolution, the rule 184 automaton applies the following rule to each of the cells in the array, simultaneously for all cells, to determine the new state of the cell, see Table 1.

Table 1. Definition of Rule 184

| Current pattern | 111 | 110 | 101 | 100 | 011 | 010 | 001 | 000 |
|---|---|---|---|---|---|---|---|---|
| New state for center cell | 1 | 0 | 1 | 1 | 1 | 0 | 0 | 0 |

Traffic flow, in mathematics and engineering, is the study of interactions between vehicles, drivers, and infrastructure (including highways, signage, and traffic control devices), with the aim of understanding and developing an optimal road network with efficient movement of traffic and minimal traffic congestion problems. [9]

Attempts to produce a mathematical theory of traffic flow date back to the 1950s, but have so far failed to produce a satisfactory general theory that can be consistently applied to real flow conditions. Current traffic models use a mixture of empirical and theoretical techniques.

Traffic phenomena are complex and nonlinear, depending on the interactions of a large number of vehicles. Due to the individual reactions of human drivers, vehicles do not interact simply following the laws of mechanics, but rather show phenomena of cluster formation and shock wave propagation, both forward and backward, depending on vehicle density in a given area. [10] In a free flowing network, traffic flow theory refers to the traffic stream variables of speed, flow, and concentration.

## 3 Defining the Centralized Control System

In this section we define the centralized control system using cellular automata for traffic flow control.

An *open-loop controller*, also called a non-feedback controller, is a type of controller which computes its input into a system using only the current state and its model of the system. A characteristic of the open-loop controller is that it does not use feedback to determine if its input has achieved the desired goal. This means that the system does not observe the output of the processes that it is controlling.

Open-loop control is useful for well-defined systems where the relationship between input and the resultant state can be modeled by a mathematical formula. This is the main reason that the cellular automaton rule 184 can be considered as a control system, in particular is an open-loop control. An open-loop controller is often used in simple processes because of its simplicity and low-cost, especially in systems where feedback is not critical. [3]

In this case, the system consists of two modules: *Forward module* and *Stop module*. Each of these modules is completely determined by a rule defined in the automaton. Forward and stop are the main actions of the vehicles at a time $t_i$, $i = 1, \ldots, n$ where $n$ is the number of evolutions in the automaton. In one evolution we have one state of the control system. This implies that the system can be observable and controllable.

Observability, in control theory, is a measure for how well internal states of a system can be inferred by knowledge of its external outputs. The observability and controllability of a system are mathematical duals. Formally, a system is said to be observable if, for any possible sequence of state and control vectors, the current state can be determined in finite time using only the outputs (this definition is slanted towards the state space representation). Less formally, this means that from the system's outputs it is possible to determine the behaviour of the entire system. Thus, the evolution of the cellular automaton is observable. [11][12]

Controllability is an important property of a control system, and the controllability property plays a crucial role in many control problems, such as stabilization of unstable systems by feedback, or optimal control. Roughly, the concept of controllability denotes the ability to move a system around in its entire configuration space using only certain admissible manipulations. The exact definition varies slightly within the framework or the type of models applied. For this case, we have analized the controllability in the behavioural framework. That is, the state of a system, which is a collection of system's variables values, completely describes the system at any given time. In particular, no information on the past of a system will help in predicting the future, if the states at the present time are known. Thus state controllability is usually taken to mean that it is possible - by admissible inputs - to steer the states from any initial value to any final value within some time window in the evolution of the automata. [11][12]

The centralized control system is defined when the cellular automata for traffic flow control is implemented in more than one dimension, in this case there are more than one track in the road. The rule for each track is the same rule 184. In this way we propose to implement a centralized control system for vehicular traffic. This system is also controllable and observable.

## 4  Conclusions

Computer science studies computable processes and information processing systems. Computational Intelligence studies problems for which there are no effective algorithms, either because it is not possible to formulate them or because they are NP-hard and thus not effective in real life applications.

Computational Intelligence research does not reject statistical methods, but often gives a complementary view and it is further closely associated with soft computing and cybernetics. A good part of its research is concerned with low-level cognitive functions: perception, signal analysis and control.

The control systems are dynamic systems that can be modeled using cellular automata. Cellular automata are an important tool for modeling and simulation of systems engineering and other research areas. The cellular automata have the ability to represent complex behaviors from a simple dynamic.

They are also studied as mathematical objects because of the intrinsic interest on the formal aspects of their behavior. It is also very important to show the changes in each state that defines the evolution patterns of behavior of the system and they are self-organizing.

## References

1. Schiff, J.L.: Cellular Automata: A Discrete View of the World. Wiley & Sons, Inc., Chichester
2. Chopard, B., Droz, M.: Cellular Automata Modeling of Physical Systems. Cambridge University Press, Cambridge (1998)
3. Kuo, B.C.: Automatic Control Systems, 6th edn. Prentice-Hall, Englewood Cliffs (1991)

4. Chandrasekharan, P.C.: Robust Control of Linear Dynamical Systems. Academic Press, London (1996)
5. Ogata, K.: Modern Control Engineering. Prentice-Hall, Englewood Cliffs (1993)
6. Wolf, D.E., Schreckenberg, M., Bachem, A. (eds.): Workshop in Traffic and Granular Flow. World Scientific, Singapore (1996)
7. Wolfram, S.: Theory and Applications of Cellular Automata. World Scientific, Singapore (1986)
8. Biham, O., Middleton, A.A., Levine, D.: Self-organization and a dynamic transition in traffic-flow models. Physical Review A 46, R6124–R6127 (1992)
9. Lieu, H.: Traffic-Flow Theory. Public Roads, US Dept. of Transportation 62(4) (January/February 1999)
10. May, A.: Traffic Flow Fundamentals. Prentice Hall, Englewood Cliffs (1990)
11. Dorf, R.C., Bishop, R.H.: Modern Control Systems. Addison-Wesley, Reading (1998)
12. D'azzo, J.J., Houpis, C.H.: Linear Control System Analysis and Design. McGraw-Hill, New York (1995)
13. Phillips, C.L., Harbor, R.D.: Feedback Control Systemas. Prentice-Hall, Englewood Cliffs (1988)

# Intelligent Methods and Models in Transportation

Mª Belén Vaquerizo García

Languages and Systems Area
Burgos University, Burgos
belvagar@ubu.es

**Abstract.** This work analyzes many different models and algorithms in the literature for the optimization of routes and frequencies of buses, necessary in the framework of the support tools development to take decisions for the collective urban public transportation systems design. The problem is NP-hard, for which diverse heuristic procedures to resolve it have been proposed in the literature. The methods that pretend to be more applicable are those that permit interactivity.

The main purpose of this work is to propose an efficient method for optimizing bus routes and their frequences considering heterogeneous vehicles, so that the final user obtains a feasible solution in a reasonable computation time. The optimization method proposed in this paper take into account a multi-objective function under diverse restrictions and an optimizer at two levels using two metaheuristic algorithms. This work presents a model based on two Genetic Algorithms.

**Keywords:** Collective Urban Transport, Routes Generator, Combinatorial Complexity, Evolutionary Algorithms, Genetic Algorithms, Multi-objective.

## 1 Introduction

Planning for an urban transport system involves determining a collective plan of routes, frequencies, schedules, assignment of coaches and staff, as best as possible.

The system, based on input data and a large number of restrictions previously established, must determine, in an effective way, an optimal design of routes, which provide a fixed level of service and cover the demands required respecting the number of buses available. The richness and difficulty of this type of problem, has made the vehicle routing an area of intense investigation.

This problem has been largely studied because of the importance of mobility in logistic field. Many different variants of this problem have been formulated to provide a suitable application to a variety of real-world cases, with the development of advanced logistic systems and optimization tools. There are different features that characterize the different variants of this problem.

The problems of allocation of fleet and personnel have been studied and are modelled as classical problems of combinatorial optimization, linear programming, and in many cases are solved in an exact way. The problem of design and optimization of routes and frequencies, by contrast, has been less studied and it is considered as an NP-hard problem. Because the problem is NP-hard to solve it, several heuristic procedures have been proposed in the literature.

The problem is addressed with an approach to combinatorial optimization, which involves in a first formulation of an optimization model and its subsequent resolution algorithm.

It should be noted that the evaluation of the solution involves calculating the objective function of the model and, moreover, the method of resolution must have a compromise between efficiency, measured in time of execution, and quality of the resulting solution. For this reason, when a routing problem is often addressed through the formulation of models whose resolution can be treated using algorithms to optimally solve instances of reasonable size in time.

This work makes a comprehensive analysis of the main models and methods or algorithms that have been used in the literature to address this problem in order to identify similarities and differences between them. Besides, it is possible to study their affinity and opportunities that may present with regard to this case of study.

This paper is organized as follow: In the following paragraph is entered to make a brief overview on the main features of the models and algorithms for optimization of routes and bus frequencies referenced in the literature. For greater simplicity, these features are presented in parameterized variables and are shared by the various proposals and their different approaches or models derived from them and explained in subsequent paragraphs. Finally, conclusions and future work are presented.

## 2 Models and Algorithms in Optimization of Bus Routes and Frequencies

In general, the optimization of an urban transport system poses collective goals such as: maximize service quality, in terms of minimizing travel times and waiting times, and maximize the benefit of the companies.

The main component that is characteristic of each model is its design, embodied primarily in its objective function. In particular the objective function reflect the interests of users, in this case the passengers, and operators, in this case the transport companies. Then it should be noted that the models presented in this section, in general, seek to maximize the level of service, minimizing resource use, under certain restrictions.

These objectives are generally contradictory, since it implies an improvement in a detriment to other, it is usually required to define a balance between these objectives.

Moreover, it should be noted that the relative importance of components of the objective function is a political decision, therefore, in practice it will usually be defined by regulatory agencies of the system. With regard to it, this paper shows the approach of each of these models and methods and algorithms [8], but the solution proposed in this paper is different because it solves the routes and the buses together, it extends the use of the genetic algorithm and improves some of its operations. For their study the most basic model is taken as a reference, considering similarities of it with the other models and methods.

For the analysis of different models, is necessary to define the characteristics of the problem to be studied and shared by all of them. It is shown in Table 1:

**Table 1.** Characteristics of the problem to be studied

| Variable | Description |
|---|---|
| n | Number of network nodes |
| $d_{ij}$ | Demand between the nodes $i$ and $j$ |
| $t_{ij}$ | travel time between $i$ and $j$ (waiting time and transfer time) |
| $t_k$ | Total travel time of route $k$ |
| $N_k$ | Number of buses operating on route $k$, $N_k = f_k t_k$ |
| $f_k$ | Frequency of buses operating on route $k$ |
| $f_{min}$ | Minimum frequency of buses allowed for the entire bus route |
| c | Available fleet size (number of buses per hour) |
| $LF_K$ | Load factor on the route $k$ |
| $(Q_k)_{max}$ | Maximum flow for each arc in the path $k$ |
| CAP | Number of seats on buses |
| $LF_{max}$ | Maximum load factor allowed |
| R | Set of routes for a given solution |
| C1, C2 | Conversion factors and relative weights of the terms of the objective function |
| $PH_{ij}$ | Number of passengers per hour between the nodes $i$ and $j$ (for the travel time of passengers) |
| $WH_{ij}$ | Waiting time for passengers between the nodes $i$ and $j$ |
| $EH_r$ | Empty travel time, which reflects the use of buses |
| a1, a2, a3 | Weights reflecting the relative importance of the terms of the function |

Moreover, it should be noted that the algorithms for optimization of routes and frequencies of buses that use these models are based on mathematical programming models solved with approximate methods, heuristics and metaheuristics.

## 3 Minimizing the Total Time for Transfer Passengers and Fleet Size Required

Through this general pattern that can be used as the basis of the following as outlined in subsequent paragraphs, which provide algorithms for optimization of routes and frequencies of buses [1]. The main aspects of the problem are taken into account, as well as a variety of parameters and constraints (load factor, etc.) [2]. It is flexible because it allows the incorporation of knowledge of users. In this regard, for example, restrictions on minimum cover ratio of demand based on free transfers or travel with at least one transfer can be added when applying a method of resolution.

In this model the components of the objective function are expressed in different units, forcing to the use of conversion factors, as it is shown in equations (1), (2), (3) and (4).

$$\text{Min } C1 \sum_{i=1}^{n} \sum_{j=1}^{n} d_{ij} t_{ij} + C2 \sum_{k \in R} f_k t_k \quad (1)$$

$$f_k \geq f_{min} \quad \forall k \in R \quad (2)$$

$$LF_k = \frac{(Q_k)_{max}}{f_k CAP} \leq LF_{max} \quad \forall k \in R \qquad (3)$$

$$\sum_{k \in R} N_k = \sum_{k \in R} f_k t_k \leq W \qquad (4)$$

**Method Used:** The proposed methodology operates on the basis of the generation, evaluation and improvement of routes. Initially generates a set of routes given the origin-destination matrix as the main guide and found the two shortest paths between a subset of *m* pairs of nodes of high demand, as seen by decreasing its value. It's necessary to specify the demand that can be uncovered is specified. Additional nodes are inserted in the initial skeleton of routes, according to pre-established rules. The generation procedure is repeated, varying parameters, giving solutions to different compromises between objectives. The main rule to assign the demand is the criterion of minimizing transfers. According to this, for each pair (*i,j*) node checks whether it is possible to travel without transfers, if not possible, alternatives to travel with 1 or 2 transfers are contemplated. Furthermore, the allocation of passenger flows in each arc of the network, and identifies valid frequencies that meet the value of the load factor set. This procedure is repeated until convergence (accepted difference between input and output frequencies of the algorithm). Moreover, the improvement of routes it operates on two distinct levels:

- Coverage of the system, discontinued service at low load of passengers, or with very short routes.
- And, the structure of the routes by combining or dividing routes.

Below are the different variations on this model are:

### 3.1 First Proposal

The first proposal uses coordinated services planning multimodal transport in heterogeneous fleet mode.

Used as the base heuristic procedures, adding the concept of transfer center (business and employment) [12]. A transfer center is detected based on data production and attraction of trips, and taking into account descriptive metrics of nodes, computed by the evaluation procedure of routes, or manually. Once identified the stops, the routes are constructed and considered. For routes passing through the centers, the frequencies are determined as a multiple of base frequency, to enable coordination between pathways that share transfer centers.

### 3.2 Second Proposal

In this case, it is necessary an algorithm to generate an initial set of routes based on the shortest path between any pair of nodes and alternative paths.

The routes are checked if they do not comply with certain restrictions (eg, minimum length) and stored as a set of candidate routes [10]. Genetic algorithms are used to select subsets of the set of candidate routes, and is a contribution in the use of metaheuristics in solving the problem.

## 3.3 Third Proposal

This proposal implies the use of algorithms for the calculation of shortest paths between any pair of nodes in the network, assign demand to check routes and restrictions on minimum and maximum passenger flow arcs.

This procedure identifies the subset of nodes that participate in the process of generating routes. In the generation of initial solutions, it is considered a single objective, to minimize travel times for passengers [11]. $k$ paths are generated between each pair of nodes of high-flow ($k$ given by the user) and used genetic algorithms to select from among any pair of $k$ nodes. In a second phase will determine the optimal frequency for the solution found in the previous phase. Again using genetic algorithms, where the objective function now incorporates the objectives of the operator, in the form of fleet costs and waiting times at the cost of the user. The main parameter that controls this process is the load factor of buses.

## 3.4 Fourth Proposal

The last proposal requires the use of Genetic Algorithms.

The proposed methodology requires an initial set of routes (the current collective urban transport system) to be improved. Using genetic algorithms, where the population is pre-cardinality, and each gene corresponds to a line, its value is an allelic pair, the first component indicates the state of the road in that configuration (active or not active) and the second value to their frequency [4]. The approach is similar to that used by the second variant shown above. The particularity of this work is that it uses a neural network to evaluate the objective function. The training of the network is done off-line based on a number of test cases, where each is an allocation process and multi-criteria analysis to determine the value of the objective function.

# 4 Proposal for Multi-objective Optimization Problem

This model is similar to that proposed above, but is formulated as a multi-objective optimization problem [7] with two objective functions, as they are shown next in equation (5) and (6).

$$\text{Min } Z1 = a1 \sum_{i,j \in N} PH_{ij} + a2 \sum_{i,j \in N} WH_{ij} + a3 \sum_{r \in R} EH_r \tag{5}$$

$$\text{Min } Z2 = W \tag{6}$$

**Method Used:** It resolves the problems of designing routes and schedules simultaneously, based on the model seen in the previous section, nonlinear mathematical programming with mixed variables, multiple objectives (minimization of travel time and minimizing the size of the fleet) [5]. The model is solved in three phases:

- First, we generate several sets of solutions not dominated by solving a joint problem of coverage.

- Below is an allocation process (not described), which determines the frequencies. For the exploration of alternative solutions using a search method that attempts to avoid local solutions already encountered, so not to start cycles.
- Finally, assess and select the most appropriate alternative, using an adapted method of "programming" for multi-objective optimization.

Its main contributions are: The formal treatment of the problem (by reducing some classic problems as subproblems to the set covering), and the method proposed for identifying non-dominated solutions.

## 5  Other Proposals

A first proposal is about the use of conversion rates of all components of the objective function: In this model, the formulation is similar to that proposed by the first method above. This model allows calculate frequency of routes, but requires the use of conversion rates to the same unit (cost / time) of all components of the objective function [9]. The objective function is evaluated through an explicit formulation that includes travel time and waiting for passengers and the cost of operating the fleet. The frequencies are determined by minimizing the value of the objective function. The last proposal is about the use of a method using Logit by calculating each line utilities for each origin-destination pair $(i,j)$: This model differs from all previous specification of the system components. It proposes an alternative allocation model, which uses the method by calculating logit utility of each line for each origin-destination pair $(i,j)$ [6]. Not dealt with issues such as determining the frequency and size of fleet. It requires the use of conversion factors and subjective values of time.

## 6  Proposal in this Paper: Multi-objective Function. Optimizer in Two Levels. Genetic Algorithms

The objective is constructing a tool to support decision making and to complement the knowledge and experience with quantitative elements. The objective is to provide a quality service, previously fixed, with minimal cost to passengers who are at different stops in the different routes. This requires to develop a system of routes and allocation of buses which is optimal in economic terms. Its development is done through a set of procedures or strategies are trying to be simple and effective.

Firstly, this model proposed to consider a multi-objective function, which consists on minimizing the number of buses and minimizing times (waiting time at stops and travel times).

Secondly, this model proposed to consider an optimizer at two levels: selection of routes and assignment of bus to them, because the solution will consist of routes and bus routes.

Besides, the complexity of a real problem of this type has the disadvantage of the many parameters that must be taken into consideration and with the objective that the solution is optimal and in real time. By this, the problem is guided through of a development process into two distinct levels, one to manage the routes and other to manage the allocation of buses on those routes, as it is shown in the next Figure.

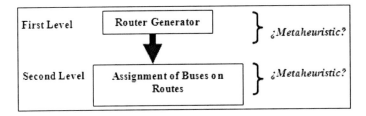

**Fig. 1.** Resolution proposed for Planning of an Urban Public Transport

Therefore, the development of a two-level optimizer for solving the proposed problem, constructs a complete solution, comprising a set of routes and a set of buses assigned to them. By this way, the construction of this solution is more refined, to ensure optimality, in two levels, each of them specialized in securing a part of the solution. The joint application of more than one metaheuristic in solving problems of great complexity, leads to build a solution with high security to be optimal.

It has been proposed to apply at each level a Genetic Algorithm, because there are many jobs where this algorithm has proved its effectiveness.

Finally, based on the frequencies obtained for these lines, the user or expert decision-maker may set hours of operation of buses in the city.

Therefore, the process adds a further complexity in the way of solving the problem, but thinks it can ensure greater reliability in the solution thus obtained is optimal. In the next figure, an experimental Session shows the performance of the optimization method proposed in this research:

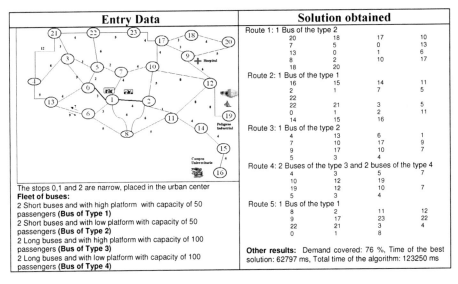

**Fig. 2.** An experimental session of the model proposed in this paper

For a network and certain information of entry there has been achieved an ideal solution formed by a set of routes and the assignment of buses on them, and taking into account that a set of restrictions were satisfied, this solution is, finally, considered as satisfactory to solve this problem.

# 7 Conclusions

As a summary of the models tested, it should be noted that all of them have a structure with similar characteristics in terms of decision variables, objective function, constraints, and so on. We also have some structural differences in terms of its development in one or two phases, on a single objective or multi-objective regarding the allocation model used.

Finally, it deserves emphasis that these algorithms have the advantage of providing a degree of interactivity to set some parameters and restrictions, are also flexible for its modularity, and allow planning in both the medium and long term [3]. Its main limitation is that it proposes a systematic way of changing the parameters to generate different solutions. There is a trend towards the use of Genetic Algorithms, similar to that occurring in other areas of combinatorial optimization. In contrast with this models analyzed, this paper has proposed an optimization method based in a multi-objective function, an optimizer at two levels and two Genetic Algorithms.

Future work includes using different fuzzy dominance approaches that should be tested to better fit the longest path better. Moreover, this algorithm can be improved with different local search algorithms, to improve the quality of solutions.

## Acknowledgements

This research has been partially supported through projects BU006A08 of the JCyL. The authors would also like to thank the vehicle interior manufacturer, Grupo Antolin Ingenieria, S.A., within the framework of the project MAGNO2008 – 1028 - CENIT Project funded by the Spanish Ministry.

## References

1. Axhausen, K.W., Smith, R.L.: Evaluation of Heuristic Transit Network Optimization Algorithms. Transportation Research Record 976 (1984)
2. Baaj, M.H., Mahmassani, H.S.: An AI-Based Approach for Transit Route System Planning and Design. Journal of Advanced Transportation 25(2) (1991)
3. Baaj, M.H., Mahmassani, H.S.: Hybrid route generation heuristic algorithm for the design of transit networks. Transportation Research 3C(1) (1995)
4. Caramia, M., Carotenuto, P., Confessore, G.: Metaheuristics techniques in bus network Optimization. In: Nectar Conference No 6 European Strategie. In: The Globalising Markets; Transport Innovations, Competitiveness and Sustainability in the Information Age, Helsinki, Finlandia (2001)
5. Ceder, A., Israeli, Y.: User and Operator Perspectives in Transit Network Design. Transportation Research Record 1623 (1998)

6. Gruttner, E., Pinninghoff, M.A., Tudela, A., Díaz, H.: Recorridos Optimos de Líneas de Transporte Público Usando Algoritmos Genéticos. Jornadas Chilenas de Computación, Copiapó, Chile (2002)
7. Israeli, Y.Y., Ceder, A.: Transit Route Design Using Scheduling and Multi-objective Programming Techniques. In: Computer-Aided Transit Scheduling, Julio de 1993, Lisboa, Portugal (1993)
8. Mauttone, A., Urquhart, M.: Optimización multiobjetivo de recorridos y frecuencias en el transporte público urbano colectivo. XIII Congreso de Ingeniería de Transporte, Universidad Pontificia Católica de Chile (2007)
9. Ngamchai, S., Lovell, D.J.: Optimal Time Transfer in Bus Transit Route Network Design Using a Genetic Algorithm. Computer-Aided Scheduling of Public Transport, Berlin, Alemania (2000)
10. Pattnaik, S.B., Mohan, S., Tom, V.M.: Urban Bus Transit Route Network Design Using Genetic Algorithm. Journal of Transportation Engineering 124(4) (1998)
11. Rao, K.V., Krishna, M.S., Dhingra, S.L.: Public Transport Routing And Scheduling Using Genetic Algorithms. Computer-Aided Scheduling of Public Transport, Berlin, Alemania (2000)
12. Shih, M.C., Mahmassani, H.S., Baaj, M.H.: Planning and Design Model for Transit (1998)

# Knowledge Based Expert System for PID Controller Tuning under Hazardous Operating Conditions

Héctor Alaiz[1], José Luis Calvo[2], Javier Alfonso[1], Ángel Alonso[1], and Ramón Ferreiro[2]

[1] Universidad de León, Edificio tecnológico, Campus de Vegazana, s/n
24071, León, Spain
{hector.moreton,javier.alfonso,angel.alonso}@unileon.es
[2] Universidad de A Coruña, Escuela Politécnica, avda. 19 de febrero, s/n,
15403, Ferrol, A Coruña, Spain
{jlcalvo,ferreiro}@udc.es

**Abstract.** The goal of this document is the exposition of a conceptual model developed and implemented like knowledge based expert system for a PID control of industrial process, when a system operates under hazardous conditions. Thanks to features of PID regulators and the expert system implemented, it is possible to stabilize a dangerous industrial process quickly if the adequate parameters of the PID regulator are calculated in a reasonable time. In this way the system can work without danger, when the communication with the plant has been lost without requiring the intervention of a human operator.

**Keywords:** Knowledge engineering, Expert system, Hazardous operating conditions, PID tuning.

## 1 Introduction

Nowadays there are many complex and critical process controlled and supervised remotely [1] [2]. When a problem in the communications between the operator and the plant appears or the system operates under hazardous conditions cause by external facts, nevertheless it is possible use a PID regulator to control this process [3].

PID control is the adjustment of the parameters that it incorporates. Above all in its topology [4] [5], as a consequence of the investigations carried out in the area, the contributions made by specialists have been many, existing among them many methods to obtain the parameters that define this regulator, achieved through different ways, and working conditions pertaining to the plant being controlled. It must be highlighted that the methods developed to obtain the terms which in occasions are empiric if they are always directed to optimize defined specifications; the negative thing is that frequently when some are improved others get worse.

With the goal to define a tool that takes the control of a process in extreme situations, tools based in techniques provided for a discipline called Knowledge Engineering have been utilized. There are similar examples about application of

Artificial Intelligent techniques successfully as [6] [7] [8] [9]. Knowledge Based Systems are used in:

- Permanence: Unlike a human expert, a knowledge based system does not grow old, and so it does not suffer loss of faculties with the pass of time.
- Duplication: Once a knowledge based system is programmed we can duplicate countless times, which reduces the costs.
- Fast: A knowledge based system can obtain information from a data base and can make numeric calculations quicker than any human being.
- Low cost: Although the initial cost can be high, thanks to the duplication capacity the final cost is low.
- Dangerous environments: A knowledge based system can work in dangerous or harmful environments for the human being.
- Reliability: A knowledge based system is not affected by external conditions, a human being yes (tiredness, pressure, etc).
- Reasoning explanation: It helps justify the exits in the case of problematic or critical domain. This quality can be employed to train personnel not qualified in the area of the application.

In accordance with what has been said, the development of a PID conceptual model is described in this document to obtain the parameters of a regulator PID with the empirical adjustment method in an open loop; feasible in the great majority of cases in which such method is applicable. The model has been developed for six groups of different expressions (formulas) with highly satisfactory results, and of course expandable to more following the same methodology.

## 2 PID Controller Conceptual Modeling

The conceptual model of a domain consists in the strictest organization possible of knowledge from the perspective of the human brain. In this sense for the domain that is being dealt with in this case study, a general summarized model is proposed and shown in figure 1.

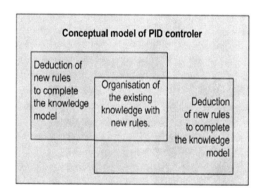

**Fig. 1.** General schema summarized from the conceptual model

As can be observed it is divided in three blocks:

- Organization of the existing rules: In this block the aim is to organise the existing rules of the types of expressions, scopes of application range, change criteria in the load disturbance or follow up of the set point control criterion, etc.
- Organization of existing knowledge with new rules: This block is the meeting point between the other two, and it aims to organise the existing knowledge in an adequate way for which it will be necessary to create new rules.
- Deduction of new rules to complete the knowledge model: In this part it has been detected the necessity to deduce new rules to make a complete knowledge model, from the own system and the desired specifications, to the final derivation of the parameters of the controller in a reasoned way.

## 3  Deduction of the Rules

As has been commented in the general summarized diagram of knowledge, it is necessary to draw new rules to complete the model of knowledge: In this part the need to do a model of complete knowledge has been detected, from the system itself and the specifications desired, up to the final retrieval of the parameters of the controller in a logical form. In this sense there are two examples in which the two possibilities of deduction of the rules are clarified.

It is pretended in this case to minimize the response time obtained, to regulate a system in which it favours the load disturbance criterion. For this in the first place, the systems of the Benchmark are arranged from less to more value of the relation L/T, as is shown in figure 17.

Following we can see illustrated and arranged from less to more value of response time obtained for the expressions to load disturbance criterion (figure 18), indicating in the graph also the expressions used in each case.

In the graph it is observed that in all cases except in three is employed the Ziegler-Nichols expressions. But contrasting the graphs 17 and 18, those systems have a relation L/T similar and elevated, being all together in the end of the graph 17. And so, two rules can be established to regulate a system for load disturbance criterion in which the time of response is improved:

- If $L/T \geq 0.6763$ expression to apply is CHR 0% Mp for load disturbance.
- If $L/T < 0.6763\ 6763$ expression to apply is Ziegler&Nichols.

This rule 2 as can be observed in figures 13 and 16; it is applied when the transfer function is not known, also in cases where it is known but does not adapt to any of the contemplated systems in the Benchmark. From it at the same time, a classification is going to be carried out, which will become three new rules.

- Rule rg. 2.1- Groups with expressions for load disturbance.
- Rule rg. 2.2- Groups with expressions setpoint control.
- Rule rg.2.3- Groups with expressions for both criteria.

To create the groups with general characteristics, the same as the previous case, the different systems are organised from less to more value with reference to L/T. In this case, it is done in a chart, because the purpose is to have generic groups in all the specifications. If for instance the case for rule rg 2.1 in which the systems are put together to follow the load disturbance criterion, then refer to chart 2.

**Table 1.** Groups rules 2.1 for load disturbance

| System | Minimum Tr | Minimum Ts | Minimum Mp | Minimum Tp |
|---|---|---|---|---|
| F  | 0,16 (Z&N)         | 1,64 (Z&N)         | 47% (CHR 0% Mp)  | 0,44 (Z&N)         |
| B1 | 0,08 (Z&N)         | 1,01 (CHR 0% Mp)   | 45% (CHR 0% Mp)  | 0,23 (Z&N)         |
| D6 | 0,53 (Z&N)         | 5,19 (CHR 20% Mp)  | 45% (CHR 0% Mp)  | 2,07 (Z&N)         |
| B2 | 0,16 (Z&N)         | 2,08 (Z&N)         | 47% (CHR 0% Mp)  | 0,44 (Z&N)         |
| A1 | 0,36 (Z&N)         | 5,29 (Z&N)         | 46% (CHR 0% Mp)  | 0,89 (Z&N)         |
| E6 | 3,6 (Z&N)          | 54,01 (CHR 0% Mp)  | 46% (CHR 0% Mp)  | 9,96 (Z&N)         |
| E5 | 1,84 (Z&N)         | 29,27 (CHR 0% Mp)  | 47% (CHR 0% Mp)  | 5,53 (Z&N)         |
| D5 | 0,5 (Z&N)          | 6,57 (CHR 0% Mp)   | 42% (CHR 0% Mp)  | 2,01 (Z&N)         |
| E4 | 0,77 (Z&N)         | 11,51 (Z&N)        | 46% (CHR 0% Mp)  | 2,52 (Z&N)         |
| A2 | 0,81 (Z&N)         | 12,87 (CHR 0% Mp)  | 45% (CHR 0% Mp)  | 2,14 (Z&N)         |
| B3 | 0,45 (Z&N)         | 5,88 (CHR 0% Mp)   | 45% (CHR 0% Mp)  | 1,26 (Z&N)         |
| C1 | 0,8 (Z&N)          | 12,81 (CHR 0% Mp)  | 45% (CHR 0% Mp)  | 2,28 (Z&N)         |
| C2 | 0,77 (Z&N)         | 13,08 (CHR 0% Mp)  | 43% (CHR 0% Mp)  | 2,39 (Z&N)         |
| A3 | 1,22 (Z&N)         | 21,15 (CHR 0% Mp)  | 41% (CHR 0% Mp)  | 3,4 (Z&N)          |
| B4 | 1,22 (Z&N)         | 21,15 (CHR 0% Mp)  | 40% (CHR 0% Mp)  | 3,4 (Z&N)          |
| D4 | 0,43 (Z&N)         | 7,76 (CHR 0% Mp)   | 39% (CHR 0% Mp)  | 1,87 (Z&N)         |
| C3 | 0,63 (Z&N)         | 13,13 (CHR 0% Mp)  | 39% (CHR 0% Mp)  | 2,59 (Z&N)         |
| C4 | 0,39 (Z&N)         | 16 (CHR 0% Mp)     | 40% (CHR 0% Mp)  | 2,72 (Z&N)         |
| E3 | 0,44 (Z&N)         | 14,05 (CHR 0% Mp)  | 32% (CHR 0% Mp)  | 1,87 (Z&N)         |
| D3 | 0,27 (Z&N)         | 11,77 (CHR 0% Mp)  | 37% (CHR 0% Mp)  | 1,59 (Z&N)         |
| D1 | 0,08 (Z&N)         | 8,19 (CHR 0% Mp)   | 46% (CHR 0% Mp)  | 1,09 (Z&N)         |
| A4 | 2,67 (Z&N)         | 68,87 (CHR 0% Mp)  | 24% (CHR 0% Mp)  | 8,86 (CHR 20% Mp)  |
| D2 | 0,15 (Z&N)         | 10 (CHR 0% Mp)     | 44% (CHR 0% Mp)  | 1,33 (Z&N)         |
| E1 | 0,18 (CHR 20% Mp)  | 10,17 (CHR 0% Mp)  | 51% (CHR 0% Mp)  | 1,53 (CHR 0% Mp)   |
| E2 | 0,41 (CHR 0% Mp)   | 13,69 (CHR 0% Mp)  | 37% (CHR 0% Mp)  | 1,75 (CHR 0% Mp)   |
| C5 | 0,37 (CHR 0% Mp)   | 90 (CHR 0% Mp)     | 102% (CHR 0% Mp) | 3,11 (CHR 0% Mp)   |

In the chart the values of the specification in each case have been indicated, alongside the expressions for obtaining the parameters used to improve this specification. Next, a division in groups is made in which the systems with groups of equal expressions are concentrated. Having this in mind, for instance systems D·, D1 and A4 with the condition that $0.6130 < L/T \leq 0.639$ (D3 to A4), and establish the following rules:

- To minimize the Response time, the Ziegler&Nichols expressions are applied.
- To minimize the Settling time, Chien, Hrones and Reswick 0% Mp.
- To optimize the Overshoot, Chien, Hrones y Reswick 0% Mp.
- To optimize the Peak time, Chien, Hrones and Reswick 20% Mp.

In spite of the systems D3 and D1 the best peak time result is obtained with ZN, if the rule for CHR of 20% a much smaller error is made than if the system A4 with ZN is regulated.

## 4 Knowledge Schema for PID Tuning

When dangerous situation appears the expert system takes the control of the situation and starts with the PID tuning following the next schema.

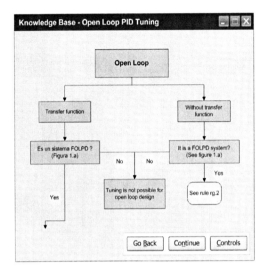

**Fig. 2.** First step of knowledge schema

In this part, the first thing to be done is to see if transfer function of the system is available. Following in both cases it is checked whether if it is a first order system with delay or if it were not the case it would not be possible to carry out the adjustment with this method. In the positive case if the transfer function is not available it concludes in the rules rg.2. If it is known the diagram of the left will be followed.

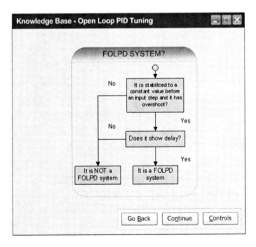

**Fig. 3.** Second step of knowledge schema

The diagram of figure 3 refers to discover if the system is a first order with delay one. For this, in first place it has to be checked if it stabilizes at a constant value with a unit step input and it is checked that there is no oscillation. If the previous is fulfilled, the next step is to check if there is a system with pure delay. With both affirmations it can be concluded that it is a system of this type, on the contrary it will not be.

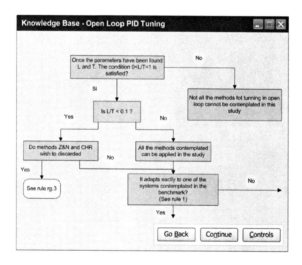

**Fig. 4.** Third step of knowledge schema

After having checked that it is a first order with delay system, and also the transfer function is known, the characteristics of the response L and T are found, as show the figure 4 and it is checked if the relation L/T is found in the application range of the expression used in the case study (between 0 and 1) if it is not so, this method of design will not be applied. If the contrary applies, L/T is checked to see if it is bigger than 0.1 and if positive it can be applied to all the expressions contemplated. If it is inferior than 0.1 the question to ask the user is if he/she wishes to discard the expressions of Ziegler-Nichols and Chien, Hrones, Reswick for not being within its scope of application range. If they are not used it will apply rule rg.3 and if they are used all methods will be taken into account as if L/T were bigger than 0.1.

After the checks of the diagram of figure 4, the diagram of figure 5 follows, the first check is to see if it adapts to the transfer function of the system being regulate adapts to some of the related in the Benchmark. If it is not the case it will follow the diagram by the right-hand area and it determines a group with general characteristics to which the function of relation L/T belongs to, that will result with rule rg.2.

If the problem system adapts exactly to one of those mentioned in the Benchmark, the system is determined and it chooses one of the three following possibilities is chosen:

- Follow a criterion of adjustment (load disturbance or set point control) and also a certain specification will be optimized. And so for instance, if what wants to be done is regulate a system before changes in the load in which the objective is to optimize the response time then it will be necessary to follow rule rg.1.1.1.
- If what is wanted is to optimize more than one specification simultaneously, rule rg.1.3. will be followed.

If what is wanted is to optimize an independent specification of the criteria of adjustment. For instance, if what is wanted is to minimize the settling time whether it is for load disturbance or set point control criteria, rule rg.1.2.2. will be followed.

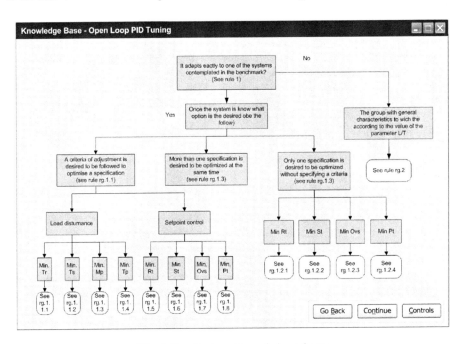

**Fig. 5.** Fourth step of knowledge schema

## 5 Conclusions

Conceptual model for open loop PID controller design showed in this document have been developed to define a procedure with the goal of controlling and stabilizing an industrial process when appear a communication problem between operation place and the plant, or another unexpected event in the operation, without requiring the intervention of an human operator. Therefore this approach is very useful for process remotely controlled.

The problem of the open loop PID design is solved with the conceptual model proposed in this work, which prime task is to choose the best expression, so we can follow the design process studying the set of rules.

When method (expression) or methods are selected, PID parameters can be calculated and the user can follow the process to obtain these parameters. The rules of the process design resolve the possible ways to get a balance between specifications to not penalize one of them. Two contributions have been extracted from the process model. First of them, has added transparency to design steeps and the second one has showed up contradictions between methods, this last problem has been resolved with our conceptual model.

When the PID parameters are calculated the controller is able to control the process and does not allow that the system causes damages due to its malfunction.

## References

1. Hamasaki, S., Yakoh, T.: Implementation and evaluation of decorators for delayed live streaming video on remote control system. In: 6th IEEE International Conference on Industrial Informatics. INDIN 2008, pp. 1220–1225 (2008)
2. Geng, X., Wu, S., Wang, J., Li, P.: An Effective Remote Control System Based on TCP/IP. In: International Conference on Advanced Computer Theory and Engineering, 2008. ICACTE 2008, pp. 425–429 (2008)
3. Itoh, M., Abe, G., Tanaka, K.: Trust in and use of automation: their dependence on occurrence patterns of malfunctions. In: IEEE International Conference on Systems, Man, and Cybernetics, 1999. IEEE SMC 1999 Conference Proceedings, vol. 3, pp. 715–720 (1999)
4. Astrom, K.J., Hagglund, T.: PID controllers: Theory, Desing and Tuning. ISA 2nd edn. Research Triangle Park (2006)
5. Feng, Y., Li., T.K.C.: PIDeasyTM and automated generation of optimal PID controllers. In: Third Asia-Pacific Conference on Control&Measurement China (1998)
6. Piramuthu, S.: Knowledge-Based Web-Enabled agents and intelligent tutoring systems. IEEE Transations on Education 48(4), 750–756 (2005)
7. Wilson, D.I.: Towards intelligence in embedded PID controllers. In: Proceedings of the Eight IASTED International Conference on Intelligent Systems and Control, Cambridge, MA (2005)
8. Epshtein, V.L.: Hypertext knowledge base for the control theory. Automation and Remote Control 61(11), 1928–1933 (2000)
9. Kaya, A., Scheib, T.J.: Tuning of PID controllers of different structures. Control Engineering 7, 62–65 (1988)

# Author Index

Abraham, Ajith   45
Agrawal, Hira   135
Alaiz, Héctor   203
Alfonso, Javier   203
Alonso, Ángel   203
Alvarez, Gonzalo   85
Álvarez, F. Hernández   163
Aztiria, Asier   19

Bacquet, Carlos   93
Bajo, Javier   77
Banerjee, Tribeni Prasad   45
Basagoiti, Rosa   19
Basile, Cataldo   117
Behrens, Clifford   135
Blasco, J.   127
Bojanić, Slobodan   171
Brito Jr., Agostinho M.   1

Caffarena, Gabriel   171
Calvo, José Luis   203
Capodieci, P.   61
Carreras, Carlos   171
Carrero, Diego   155
Castillo, Andrés   109
Castillo, Luis   109
Chen, Hao   101
Chivers, Howard   101
Choudhury, Joydeb Roy   45
Clark, John A.   101
Corchado, Emilio   143
Corchado, Juan M.   77

Das, Swagatam   45
Dasarathy, Balakrishnan   135
Decherchi, Sergio   29, 37
de Fuentes, J.M.   127
de Jesús Medel-Juárez, José   187
De Paz, Juan F.   77
Dhillon, Paramveer S.   179

Encinas, L. Hernández   163

Faggioni, Osvaldo   37
Ferreiro, Ramón   203

García, M$^a$ Belén Vaquerizo   193
Gastaldo, Paolo   37
Gómez, Juan Miguel   155

Herrero, Álvaro   143
Heywood, Malcolm I.   93

Isaza, Gustavo   109

Jiang, J.   61
Jiang, Y.   61
Jiménez-Benítez, José Alfredo   187

Lee Fook, Leslie   135
Leoncini, Alessio   29
Leoncini, Davide   37
Lioy, Antonio   117
López, Manuel   109

Madheswaran, M.   69
Medeiros, João Paulo S.   1
Menakadevi, T.   69
Milovanović, Vladimir   171
Munilla, Jorge   53

Nobles, Philip   101

Ortiz, Andrés   53

Palma-Orozco, Gisela   187
Palma-Orozco, Rosaura   187
Palomar, E.   127
Peinado, Alberto   53
Pejović, Vukašin   171
Perez-Villegas, Alejandro   85
Pinzón, Cristian I.   77
Pires, Paulo S. Motta   1
Popović, Jelena   171
Poza, M. Jesús   155
Puente, Luis   155

Redi, Judith   29
Reyes, Mario   19
Ribagorda, A.   127
Rolando, Menchaca García Felipe   9

# Author Index

Salvador, Contreras Hernández  9
Sangiacomo, Fabio  29
Santafé, Guzmán  19
Scozzi, Salvatore  117
Shaikh, Siraj A.  101
Snášel, Václav  45
Soldani, Maurizio  37
Suarez-Tangil, G.  127

Tacconi, Simone  29
Tapiador, Juan E.  101
Torrano-Gimenez, Carmen  85

Vallini, Marco  117

Zincir-Heywood, A. Nur  93
Zunino, Rodolfo  29, 37
Zurutuza, Urko  19

Breinigsville, PA USA
23 September 2009
224580BV00003B/6/P